CRC SERIES IN RADIOTRACERS IN BIOLOGY AND MEDICINE

Editor-in-Chief

Lelio G. Colombetti, Sc.D.
Loyola University
Stritch School of Medicine
Maywood, Illinois

STUDIES OF CELLULAR FUNCTION USING RADIOTRACERS
Mervyn W. Billinghurst, Ph.D.
Radiopharmacy
Health Sciences Center
Winnipeg, Manitoba, Canada

GENERAL PROCESSES OF RADIOTRACER LOCALIZATION
Leopold J. Anghileri, D.Sc.
Laboratory of Biophysics
University of Nancy
Nancy, France

RADIATION BIOLOGY
Donald Pizzarello, Ph.D.
Department of Radiology
New York University Medical Center
New York, New York

RADIOTRACERS FOR MEDICAL APPLICATIONS
Garimella V. S. Rayudu, Ph.D
Nuclear Medicine Department
Rush University Medical Center
Presbyterian - St. Luke's Hospital
Chicago, Illinois

RECEPTOR-BINDING RADIOTRACERS
William C. Eckelman, Ph.D.
Department of Radiology
George Washington University School of Medicine
Washington, D.C.

BIOLOGIC APPLICATIONS OF RADIOTRACERS
Howard J. Glenn, Ph.D.
University of Texas System Cancer Center
M.D. Anderson Hospital and Tumor Institute
Houston, Texas

BIOLOGICAL TRANSPORT OF RADIOTRACERS
Lelio G. Colombetti, Sc.D.
Loyola University
Stritch School of Medicine
Maywood, Illinois

BASIC PHYSICS
W. Earl Barnes, Ph.D.
Nuclear Medicine Service
Edward Hines, Jr., Hospital
Hines, Illinois

RADIOBIOASSAYS
Fuad S. Ashkar, M.D.
Radioassay Laboratory
Jackson Memorial Medical Center
University of Miami School of Medicine
Miami, Florida

Basic Physics of Radiotracers

Volume I

Editor

W. Earl Barnes, Ph.D.
Physicist
Nuclear Medicine Services
Edward Hines, Jr., Veterans Administration Hospital
Hines, Illinois

Editor-in-Chief
CRC Series in Radiotracers in Biology and Medicine

Lelio G. Colombetti, Sc.D.
Loyola University
Stritch School of Medicine
Maywood, Illinois

CRC Press
Taylor & Francis Group
Boca Raton London New York

CRC Press is an imprint of the
Taylor & Francis Group, an informa business

First published 1983 by CRC Press
Taylor & Francis Group
6000 Broken Sound Parkway NW, Suite
300 Boca Raton, FL 33487-2742

Reissued 2018 by CRC Press

A Library of Congress record exists under LC control number: 82014723

ISBN 13: 978-1-138-50657-2 (hbk)
ISBN 13: 978-1-315-14994-3 (ebk)

Visit the Taylor & Francis Web site at http://www.taylorandfrancis.com and the CRC Press Web site at http://www.crcpress.com

FOREWORD

This series of books on Radiotracers in Biology and Medicine is on the one hand an unbelievably expansive enterprise and on the other hand, a most noble one as well. Tools to probe biology have developed at an accelerating rate. Hevesy pioneered the application of radioisotopes to the study of chemical processes, and since that time, radioisotopic methodology has probably contributed as much as any other methodology to the analysis of the fine structure of biologic systems. Radioisotopic methodologies represent powerful tools for the determination of virtually any process of biologic interest. It should not be surprising, therefore, that any effort to encompass all aspects of radiotracer methodology is both desirable in the extreme and doomed to at least some degree of inherent failure. The current series is assuredly a success relative to the breadth of topics which range from in depth treatise of fundamental science or abstract concepts to detailed and specific applications, such as those in medicine or even to the extreme of the methodology for sacrifice of animals as part of a radiotracer distribution study. The list of contributors is as impressive as is the task, so that one can be optimistic that the endeavor is likely to be as successful as efforts of this type can be expected to be. The prospects are further enhanced by the unbounded energy of the coordinating editor. The profligate expansion of application of radioisotopic methods relate to their inherent and exquisite sensitivity, ease of quantitation, specificity, and comparative simplicity, especially with modern instrumentation and reagents, both of which are now readily and universally available. It is now possible to make biological measurements which were otherwise difficult or impossible. These measurements allow us to begin to understand processes in depth in their unaltered state so that radioisotope methodology has proved to be a powerful probe for insight into the function and perturbations of the fine structure of biologic systems. Radioisotopic methodology has provided virtually all of the information now known about the physiology and pathophysiology of several organ systems and has been used abundantly for the development of information on every organ system and kinetic pathway in the plant and animal kingdoms. We all instinctively turn to the thyroid gland and its homeostatic interrelationships as an example, and an early one at that, of the use of radioactive tracers to elaborate normal and abnormal physiology and biochemistry, but this is but one of many suitable examples. Nor is the thyroid unique in the appreciation that a very major and important residua of diagnostic and therapeutic methods of clinical importance result from an even larger number of procedures used earlier for investigative purposes and, in some instances, procedures used earlier for investigative purposes and, in some instances, advocated for clinical use. The very ease and power of radioisotopic methodology tempts one to use these techniques without sufficient knowledge, preparation or care and with the potential for resulting disastrous misinformation. There are notable research and clinical illustrations of this problem, which serve to emphasize the importance of texts such as these to which one can turn for guidance in the proper use of these powerful methods. Radioisotopic methodology has already demonstrated its potential for opening new vistas in science and medicine. This series of texts, extensive though they be, yet must be incomplete in some respects. Multiple authorship always entails the danger of nonuniformity of quality, but the quality of authorship herein assembled makes this likely to be minimal. In any event, this series undoubtedly will serve an important role in the continued application of radioisotopic methodology to the exciting and unending, yet answerable, questions in science and medicine!

Gerald L. DeNardo, M.D.
Professor of Radiology, Medicine,
Pathology and Veterinary Radiology
University of California, Davis-
Sacramento Medical School
Director, Division of Nuclear Medicine

THE EDITOR-IN-CHIEF

Lelio G. Colombetti, Sc.D., is Professor of Pharmacology at Loyola University Stritch School of Medicine in Maywood, Ill. and a member of the Nuclear Medicine Division Staff at Michael Reese Hospital and Medical Center in Chicago, Ill.

Dr. Colombetti graduated from the Litoral University in his native Argentina with a Doctor in Sciences degree (summa cum laude), and obtained two fellowships for postgraduate studies from the Georgetown University in Washington, D.C., and from the M.I.T. in Cambridge, Mass. He has published more than 150 scientific papers and is the author of several book chapters. He has presented over 300 lectures both at meetings held in the U.S. and abroad. He organized the First International Symposium on Radiopharmacology, held in Innsbruck, Austria, in May 1978. He also organized the Second International Symposium on Radiopharmacology which took place in Chicago in September, 1981, with the active participation of more than 500 scientists, representing over 30 countries. He is a founding member of the International Association of Radiopharmacology, a nonprofit organization, which congregates scientists from many disciplines interested in the biological applications of radiotracers. He was its first President (1979/1981).

Dr. Colombetti is a member of various scientific societies, including the Society of Nuclear Medicine (U.S.) and the Gesellschaft für Nuklearmedizin (Europe), and is an honorary member of the Mexican Society of Nuclear Medicine. He is also a member of the Society of Experimental Medicine and Biology, the Coblenz Society, and the Sigma Xi. He is a member of the editorial boards of the journals *Nuklearmedizin* and *Research in Clinic and Laboratory.*

PREFACE

Few fields of research draw upon a more diverse array of scientific disciplines than does that of radiotracer research in biology and medicine. To master these all would tax the abilities of a Leonardo da Vinci, and as a consequence the researcher becomes expert only in those areas most germane to his work. One discipline frequently receiving short shrift in the education of persons working in the biomedical radiotracer field is the physics of the atom and nucleus as it relates to the nature of radioactive decay and of radiation. This subject is fundamental to all radiotracer experiments and in addition forms the underpinnings of the more applied fields of nuclear instrumentation, radiochemistry, radionuclide production, and radiation dosimetry.

The opportunity to present the physics of radioactive processes in some detail and apart from topics such as instrumentation which conventionally compete with it for space is most welcome. The material is intended to give a fairly complete introduction to radiation physics to those who wish to have more than a descriptive understanding of the subject. Although it is possible to work one's way through much of the subject matter without having had any previous physics background, some prior acquaintance with modern physics is desirable. A familiarity with calculus and differential equations is also assumed.

Volume I begins with a brief description of classical physics, its extension to special relativity and quantum mechanics, and an introduction to basic atomic and nuclear concepts. A thorough discussion of atomic structure follows with emphasis on the theory of the multielectron atom, characteristic X-rays, and the Auger effect. Volume II treats the subjects of nuclear structure, nuclear decay processes, the interaction of radiation with matter, and the mathematics of radioactive decay.

Lelio G. Colombetti, Sc.D.
W. Earl Barnes, Ph.D.

THE EDITOR

W. Earl Barnes, Ph.D., is Physicist at the Veterans Hospital, Hines, Illinois.

Dr. Barnes received the B.A. degree from Harvard University in 1961 and the Ph.D. degree in nuclear physics in 1970 from Oregon State University. In addition to a teaching program in physics, he is engaged in research which applies mathematics and physics to various problems of medicine, having published approximately 40 articles in the areas of mathematical modeling of liver and bone kinetics, functional imaging of the heart and lungs, and radiation dosimetry. Currently Treasurer of the International Association of Radiopharmacology, he has been active in the organization of the First and Second International Symposiums on Radiopharmacology.

Dr. Barnes is a member of Sigma Xi, Sigma Pi Sigma, and the Society of Nuclear Medicine.

CONTRIBUTORS

Harry T. Easterday, Ph.D.
Professor of Physics
Oregon State University
Corvallis, Oregon

B. T. A. McKee, Ph.D.
Associate Professor
Department of Physics
Assistant Professor
Department of Medicine
Queen's University
Kingston, Ontario, Canada

R. R. Sharma, Ph.D.
Professor
Department of Physics
University of Illinois
Chicago, Illinois

Donald A. Walker, Ph.D.
Professor
Department of Physics
State University of New York
New Paltz, New York

TABLE OF CONTENTS

Volume I

Volume II

Chapter 1

INTRODUCTORY CONCEPTS

W. Earl Barnes

TABLE OF CONTENTS

I. INTRODUCTION

Physical theories are commonly classified as being either "classical" or "modern". The reasons for this distinction are both historical and substantive. Limited in the sophistication of their measuring instruments, early scientists proposed theories appropriate for the description of the simplest and most accessible physical phenomena, e.g., the trajectories of the planets. Because of the class of phenomena observed, certain beliefs came to underlie all classical theories with regard to the nature of time, space, matter, etc. For example, the idea was undisputed that an object has at all times both a definite position and velocity. Unfortunately, our physical intuition is easily deceived. Not until the interior of the atom and the nature of electromagnetic radiation were explored was it discovered that the concepts of classical physics are inadequate to deal with many phenomena. A reassessment of fundamental postulates led to the formulation of modern physics which, it is believed, successfully treats the behavior of all physical systems.

To gain an understanding of the rudiments of modern physics, we proceed as the early scientists did by first mastering the classical concepts that emerge from our intuitive picture of the world. Modifications of these concepts are subsequently introduced which allow a more accurate treatment of physical phenomena, particularly atomic and nuclear systems.

II. MATHEMATICS OF VECTORS

Observations in science and in everyday life often yield a number. For example, we measure the temperature of the air or the length of a table. Sometimes direction is also involved. Thus, a flock of birds is not only flying at 30 km/hr but also in a south-westerly direction. A quantity that encompasses both magnitude and direction is the **vector**, an extraordinarily useful concept in mathematics and physics. The usefulness of vectors results in part from the fact that vector formulas can express complicated spatial and numerical relationships in a concise way and without the use of coordinate axes.

Ordinary numbers or magnitudes are called **scalars**. Time, length, and mass are some physical quantities that are scalars.

Vectors are used to represent quantities that have both magnitude and direction associated with them. Velocity and force are examples of vector quantities. Vectors are usually pictured as arrows, where the length of the arrow is the magnitude of the vector and the direction in which the arrow is pointing is the direction of the vector. Two vectors are considered to be identical if they have the same magnitude and direction. In a vector diagram, it is always possible to "slide" a vector anywhere in the diagram without changing its fundamental character as long as the original magnitude and direction of the vector are preserved.

A. Vector Addition

The algebra of vectors is somewhat different from the algebra of ordinary numbers. We use boldface type to depict vectors algebraically. **Vector addition** is shown in Figure 1 where vector **A** is added to vector **B**, the result being vector **C**. Geometrically, one simply lays out vector **B** with its tail at the head of vector **A** and then draws a line from the tail of **A** to the head of **B** to form **C**. It is an easy matter to verify that it makes no difference in what order or grouping we add vectors; the sum is the same.

Vector subtraction is defined by taking the negative of a vector to be a vector of equal magnitude (length) but opposite in direction.

There is an alternative way of adding or subtracting vectors: the **method of components.** In Figure 2 we have vector **A** lying in the xy-plane. The projections of **A** on the x- and y-axes are called the x- and y-components of **A**, and these are the scalars A_x and A_y. Let us symbolize the magnitude (or length) of vector **A** by the bold letter **A**. Then the relation between A and the vector components is, by the Pythagorean theorem,

$$A = \sqrt{A_x^2 + A_y^2}$$

Also, if θ is the angle that the vector makes with the x-axis, then

$$\cos\theta = A_x/A = A_x/\sqrt{A_x^2 + A_y^2}$$

$$\sin\theta = A_y/A = A_y/\sqrt{A_x^2 + A_y^2}$$

Now, in order to add together vectors **A** and **B** to form vector **C**, we use a simple rule: the x-component of the sum of several vectors is the algebraic sum of the x-components of the vectors; and similarly for the y- and z-components. That is, if **C** = **A** + **B**, then the components of **C** are

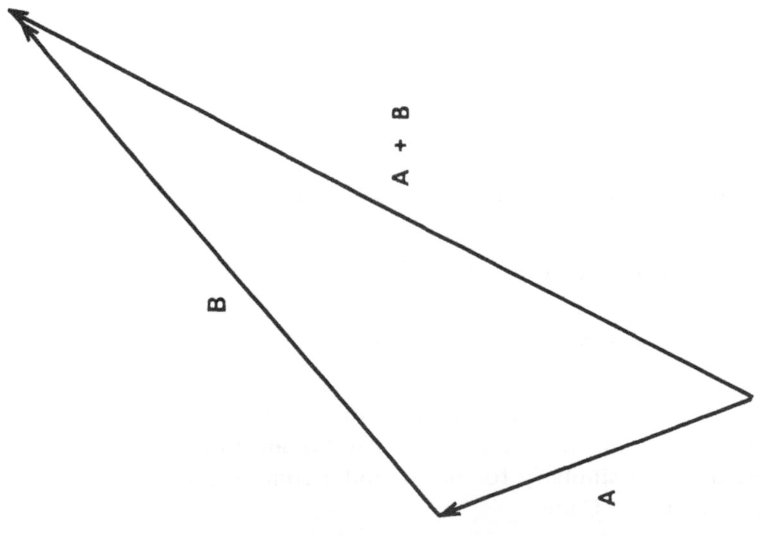

FIGURE 1. Vector addition A + B = C.

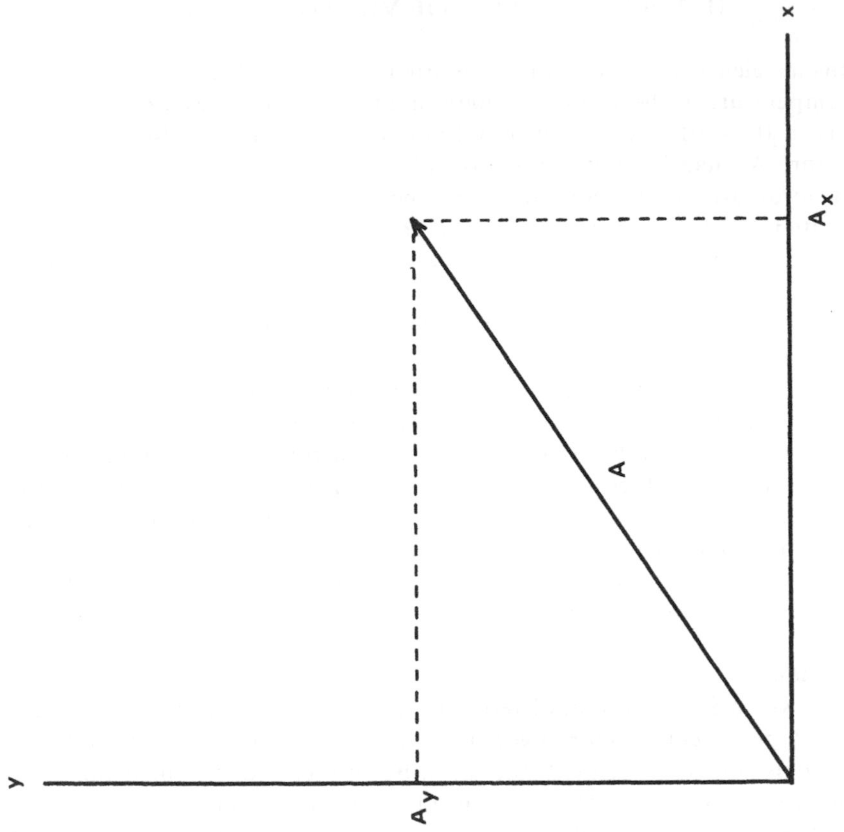

FIGURE 2. The components A_x and A_y of vector A.

$$C_x = A_x + B_x$$

$$C_y = A_y + B_y$$

$$C_z = A_z + B_z$$

The magnitude of **C** is given by:

$$C = \sqrt{C_x^2 + C_y^2 + C_z^2}$$

The angle θ that **C** makes with any coordinate axis, for example the x-axis, is found from:

$$\cos\theta = C_x/C = C_x/\sqrt{C_x^2 + C_y^2 + C_z^2}$$

Example 1. Vector Addition

An airplane is flying due east at 100 km/hr in no wind. Subsequently, it encounters a gust of wind from the south of 50 km/hr. To find the new velocity of the plane, we must add together the two velocity vectors. If **A** represents the initial velocity of the plane, **B** the velocity of the wind, and **C** the new velocity of the plane, then

$$C_{East} = A_{East} + B_{East}$$

$$= 100 + 0$$

$$= 100 \text{ km/hr}$$

$$C_{North} = A_{North} + B_{North}$$

$$= 0 + 50$$

$$= 50 \text{ km/hr}$$

$$C = \sqrt{C_{East}^2 + C_{North}^2}$$

$$= \sqrt{100^2 + 50^2}$$

$$= 111.8 \text{ km/hr}$$

The angle θ that **C** makes with the East-West direction is found from

$$\sin\theta = C_{North}/C$$

$$= 50/111.8$$

$$= 0.447$$

so that $\theta = 26.56°$.

B. Multiplication of Vectors

1. Multiplication of a Vector by a Scalar

The **multiplication of a vector by a scalar** has a simple meaning: each of the components of the vector is multiplied by the scalar. For example, the product of the scalar k and the vector **A**, written k**A**, is the new vector **B** with the components:

$$B_x = k A_x$$

$$B_y = k A_y$$

$$B_z = k A_z$$

The magnitude of the new vector **B** is simply k times the magnitude of **A**. The direction of the new vector is the same as that of the old vector if k is a positive number and is opposite in direction if k is negative.

One may represent vectors in a way that is often useful using vector components and **unit vectors**, which are vectors whose magnitude is unity. It is customary to invent the three unit vectors **i**, **j**, and **k**, which point along the x- , y- , and z-axes, respectively. Then any vector **A** can be expressed as the sum of each component multiplied by its corresponding unit vector:

$$\mathbf{A} = A_x\,\mathbf{i} + A_y\,\mathbf{j} + A_z\,\mathbf{k}$$

2. Dot Product of Two Vectors

The **dot product** of two vectors is a method of vector multiplication that results in a scalar. If the two vectors are not positioned so that they originate from the same point, it will be necessary to so position them, remembering that it is permissible to "slide" any vector as long as the direction in which it points is not changed. The dot product of vectors **A** and **B**, written **A** · **B**, is defined to be

$$\mathbf{A} \cdot \mathbf{B} = AB \cos\theta$$

where θ is the angle between **A** and **B**. When **A** and **B** are in the same direction, then $\theta = 0°$ and $\cos\theta = 1$ so that **A** · **B** = AB. If **A** and **B** happen to be at right angles to each other, then $\theta = 90°$ and $\cos\theta = 0$ so that **A** · **B** = 0.

For the unit vectors **i**, **j**, and **k** lying along the x- , y- , and z-axes, since their magnitudes are unity and they are mutually perpendicular to each other,

$$\mathbf{i}\cdot\mathbf{i} = \mathbf{j}\cdot\mathbf{j} = \mathbf{k}\cdot\mathbf{k} = 1$$

and

$$\mathbf{i}\cdot\mathbf{j} = \mathbf{i}\cdot\mathbf{k} = \mathbf{j}\cdot\mathbf{k} = 0$$

Hence, the dot product of **A** and **B** may be expressed as

$$\mathbf{A}\cdot\mathbf{B} = (A_x\,\mathbf{i} + A_y\,\mathbf{j} + A_z\,\mathbf{k}) \cdot (B_x\,\mathbf{i} + B_y\,\mathbf{j} + B_z\,\mathbf{k})$$

$$= A_x B_x + A_y B_y + A_z B_z$$

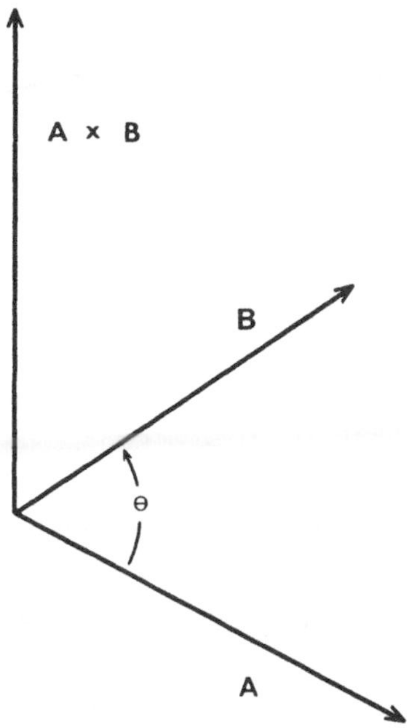

FIGURE 3. The vector cross product $\mathbf{C} = \mathbf{A} \times \mathbf{B}$. The "right-hand rule" states that if one curls the fingers so as to rotate **A** into **B**, then the thumb points in the direction of **C**.

3. Cross Product of Two Vectors

A third method of vector multiplication is the **cross product** of two vectors in which the result is a vector. First the two vectors are positioned so that they originate from the same point. Then the cross product of vectors **A** and **B**, written $\mathbf{A} \times \mathbf{B}$, is a new vector **C** which has magnitude

$$C = A\,B \sin\theta$$

where θ is the angle between **A** and **B**. The direction of **C** is defined to be perpendicular to the plane formed by **A** and **B** and is easily identified by the "right-hand rule" as illustrated in Figure 3: coil the fingers of your right hand so as to push **A** into **B**; then your thumb will point in the direction of **C**. Notice that $\mathbf{A} \times \mathbf{B} = -\mathbf{B} \times \mathbf{A}$ so that the order of factors in a cross product is important.

C. Derivative of a Vector

Vectors may vary, for example, as a function of time t. The derivative of a vector function is defined in the following way. Suppose vector **A** is equal to $\mathbf{A}_1$ at time t_1, and equal to $\mathbf{A}_2$ at time t_2. We call Δt the time interval between t_1 and t_2, i.e., $\Delta t = t_2 - t_1$. Then the derivative of **A** with respect to t is defined to be

$$\frac{d\mathbf{A}}{dt} = \lim_{\Delta t \to 0} \frac{\mathbf{A}_2 - \mathbf{A}_1}{\Delta t} = \lim_{\Delta t \to 0} \frac{\Delta \mathbf{A}}{\Delta t}$$

where $\Delta t \to 0$ means the limiting value that is approached as Δt approaches zero. In terms of components,

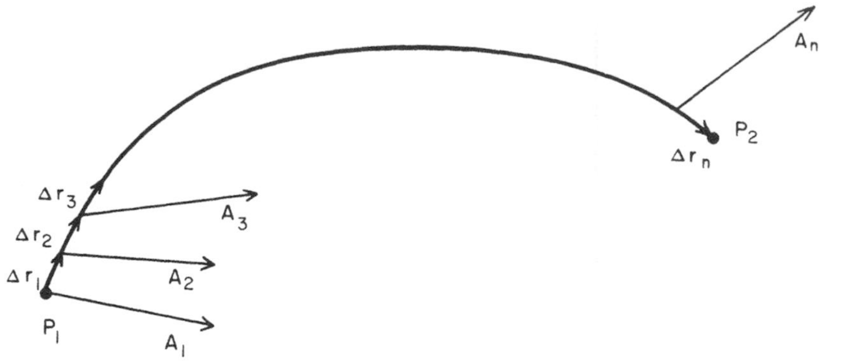

FIGURE 4. The line integral $\int^{P_2} \mathbf{A} \cdot d\mathbf{r}$ along a path from P_1 to P_2 as the sum of the dot products of the vector **A** with the line element vectors $\Delta\mathbf{r}$.

$$\mathbf{A}_2 - \mathbf{A}_1 = (A_{2x}\,\mathbf{i} + A_{2y}\,\mathbf{j} + A_{2z}\,\mathbf{k}) - (A_{1x}\,\mathbf{i} + A_{1y}\,\mathbf{j} + A_{1z}\,\mathbf{k})$$

$$= (A_{2x} - A_{1x})\,\mathbf{i} + (A_{2y} - A_{1y})\,\mathbf{j} + (A_{2z} - A_{1z})\,\mathbf{k}$$

On dividing by Δt and letting Δt approach zero, we have

$$\frac{d\mathbf{A}}{dt} = \frac{dA_x}{dt}\,\mathbf{i} + \frac{dA_y}{dt}\,\mathbf{j} + \frac{dA_z}{dt}\,\mathbf{k}$$

Hence, to differentiate a vector function one differentiates each component separately.

D. Integral of a Vector

The simple **integral** of a vector function of one variable, e.g., time, is defined in a manner analogous to that for the derivative. If the interval of time between t_1 and t_2 is divided into n equal segments Δt and if the corresponding values of the vector **A** at each of these times are $\mathbf{A}_1, \mathbf{A}_2, \cdots, \mathbf{A}_n$, then the integral of **A** is defined as the sum of the products of the vectors with the time interval, in the limit as the number of segments is made infinitely large:

$$\int_{t_1}^{t_2} \mathbf{A}\, dt = \lim_{n\to\infty} \sum_{i=1}^{n} \mathbf{A}_i\, \Delta t$$

A second type of vector integral can be defined when associated with each point in space there is a vector quantity, such as the velocity of a fluid or the force on an object. It is often useful to find the **line integral** of such a vector quantity along a curve through the space. The line integral is defined in the following way. A curve joining the points P_1 and P_2 is divided into n arc elements; and at each arc element is drawn a vector $\Delta\mathbf{r}$ connecting the beginning and end of the element as shown in Figure 4. Suppose the vector quantity **A** has values $\mathbf{A}_1, \mathbf{A}_2, \cdots, \mathbf{A}_n$ at each element of the curve. We then form the dot products of **A** with the corresponding arc element vectors $\Delta\mathbf{r}$: $\mathbf{A}_1 \cdot \Delta\mathbf{r}_1, \mathbf{A}_2 \cdot \Delta\mathbf{r}_2, \cdots, \mathbf{A}_n \cdot \Delta\mathbf{r}_n$. The summation of these dot products, in the limit as the number n of arc elements is made infinitely large, is the line integral:

$$\int_{P_1}^{P_2} \mathbf{A} \cdot d\mathbf{r} = \lim_{n\to\infty} \sum_{i=1}^{n} \mathbf{A}_i \cdot \Delta \mathbf{r}_i$$

Because the dot product is proportional to the cosine of the angle between the two vectors, the line integral will be a maximum when the curve follows the direction of the vector quantity **A**. When the curve is at right angles to **A**, the line integral vanishes.

In terms of vector components we may write

$$\mathbf{A} = A_x \mathbf{i} + A_y \mathbf{j} + A_z \mathbf{k}$$

and

$$d\mathbf{r} = dx\, \mathbf{i} + dy\, \mathbf{j} + dz\, \mathbf{k}$$

so that the line integral becomes

$$\int_{P_1}^{P_2} \mathbf{A} \cdot d\mathbf{r} = \int_{P_1}^{P_2} (A_x \mathbf{i} + A_y \mathbf{j} + A_z \mathbf{k}) \cdot (dx\, \mathbf{i} + dy\, \mathbf{j} + dz\, \mathbf{k})$$

$$= \int_{P_2}^{P_1} (A_x\, dx + A_y\, dy + A_z\, dz)$$

III. CLASSICAL PHYSICS

Based on observations of the most accessible physical phenomena, the laws of classical physics are generally valid descriptions of how ordinary-sized objects behave when moving at ordinary speeds. Mechanics, which is the study of the effects of forces on the motion of objects, is one of the oldest of sciences and the most fundamental of the physical sciences. Electromagnetism is the study of electric charges and their effects. Together, these two branches of classical physics, although inadequate to describe phenomena on the atomic scale, provide most of the concepts and mathematical machinery necessary to understand atomic and nuclear structure.

A. Mechanics

1. Particle Kinematics

Objects typically are capable of rotating or of having various internal motions. In order to avoid these complications we will deal in the main with **particles**, objects that can be treated as mathematical points. The description of the motion of particles is called particle **kinematics**.

The position of a particle can be described by specifying its coordinates in a suitable coordinate system. Equivalently, one can specify the position of a particle by using a vector to represent the distance and direction of the particle from some reference point.

a. Velocity

The **velocity** of a particle is its time rate of change of position and is a vector quantity. Suppose that at time t_1 a particle is located at point P_1 which is specified by position vector $\mathbf{R}_1$, and at time t_2 the particle is located at point P_2 specified by vector $\mathbf{R}_2$. Then the average velocity $\mathbf{v}_{av}$ of the particle is defined by

$$\mathbf{v}_{av} = \frac{\mathbf{R}_2 - \mathbf{R}_1}{t_2 - t_1}$$

The direction of $\mathbf{v}_{av}$ is from P_1 to P_2. The magnitude of the velocity vector is commonly called the **speed** of the particle and is equal to the distance between P_1 and P_2 divided by the time interval $(t_2 - t_1)$.

The closer P_2 is chosen to P_1, the more accurately we know the **instantaneous velocity** of the particle at P_1. Mathematically, we express this idea in the following way. We call

$$\Delta\mathbf{R} = \mathbf{R}_2 - \mathbf{R}_1$$

and

$$\Delta t = t_2 - t_1$$

We expect the ratio $\Delta\mathbf{R}/\Delta t$ to approach a limiting value as the time interval Δt and the vector displacement $\Delta\mathbf{R}$ become increasingly small. This limiting value, the instantaneous velocity $\mathbf{v}$, is

$$\mathbf{v} = \lim_{\Delta t \to 0} \frac{\Delta\mathbf{R}}{\Delta t}$$

$$= \frac{d\mathbf{R}}{dt}$$

b. Acceleration

Acceleration is the time rate of change of velocity. It is important to note that the velocity can vary because of a change in the speed of the particle, because of a change in direction of the particle, or both. Suppose that at time t_1 a particle is moving with instantaneous velocity $\mathbf{v}_1$ and at time t_2 with the velocity $\mathbf{v}_2$. Then the **average acceleration** of the particle is the change in velocity divided by the time interval:

$$\mathbf{a}_{av} = \frac{\mathbf{v}_2 - \mathbf{v}_1}{t_2 - t_1}$$

We can define an **instantaneous acceleration** just as we defined an instantaneous velocity. During the short time interval $\Delta t = t_2 - t_1$, the change in velocity is $\Delta\mathbf{v} = \mathbf{v}_2 - \mathbf{v}_1$. The instantaneous acceleration $\mathbf{a}$ is the limiting value of $\Delta\mathbf{v}/\Delta t$ as Δt approaches zero:

$$\mathbf{a} = \lim_{\Delta t \to 0} \frac{\Delta\mathbf{v}}{\Delta t}$$

$$= \frac{d\mathbf{v}}{dt}$$

$$= \frac{d^2\mathbf{R}}{dt^2}$$

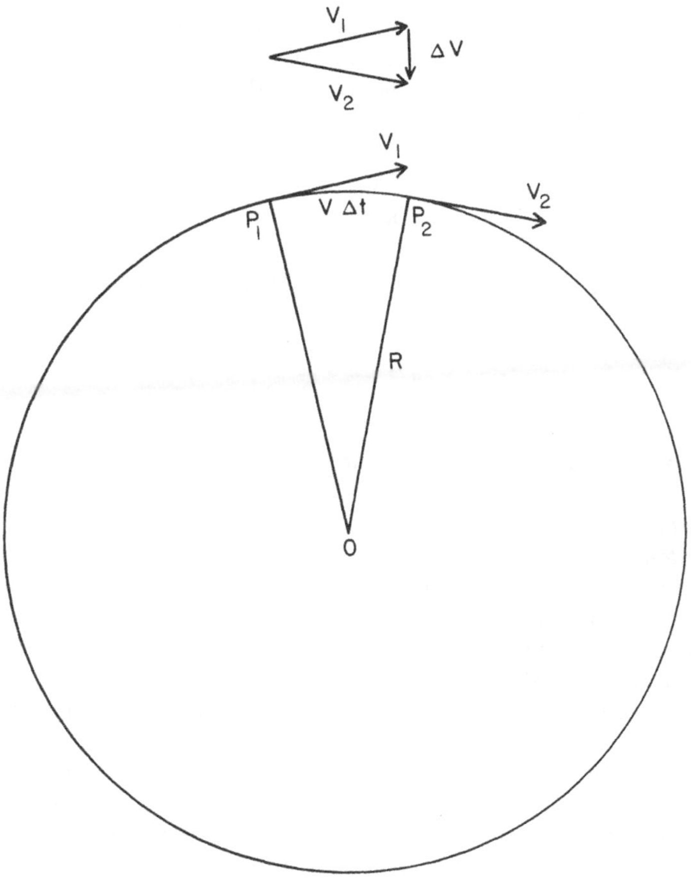

FIGURE 5. The acceleration of a particle moving in a circle of radius R with constant speed v. At two nearby points in its motion its velocity is $\mathbf{v}_1$ and $\mathbf{v}_2$, respectively. The change in its velocity is $\Delta\mathbf{v} = \mathbf{v}_2 - \mathbf{v}_1$ which points toward the center of the circle.

Example 2. Particle Moving Without Acceleration

Consider a particle that has no acceleration. Since velocity has both a magnitude and direction associated with it, a nonaccelerating particle must neither change its speed nor direction of motion. Hence, a nonaccelerating particle moves in a straight line at constant speed.

Example 3. Particle Moving in Circle with Constant Speed

Consider an object whose speed does not change but whose direction of motion is constantly changing. A particle moving with constant speed around a circular path is such an example. Suppose P_1 is the position of the particle at time t_1 and P_2 is the position of the particle at time t_2, as shown in Figure 5. The instantaneous velocity of the particle is always a tangent to the circle at the position of the particle. To find the acceleration **a**, it is necessary to compute the change in the velocity $\Delta\mathbf{v} = \mathbf{v}_2 - \mathbf{v}_1$ divided by the time interval $\Delta t = t_2 - t_1$:

$$\mathbf{a} = \frac{\mathbf{v}_2 - \mathbf{v}_1}{t_2 - t_1} = \frac{\Delta\mathbf{v}}{\Delta t}$$

Now if we draw $\mathbf{v}_1$ and $\mathbf{v}_2$ so that they originate from the same point, we form an isosceles triangle that is similar to the triangle OP_1P_2 by virtue of the fact that $\mathbf{v}_1$ is perpendicular to OP_1 and $\mathbf{v}_2$ is perpendicular to OP_2. Hence, the ratio of the base to the side is equal for both:

$$\frac{v\,\Delta t}{R} = \frac{\Delta v}{v}$$

where R is the radius of the circle. So the magnitude of the acceleration **a** is

$$a = \frac{\Delta v}{\Delta t} = \frac{v^2}{R}$$

The direction of **a** is toward the center of the circle.

2. Particle Dynamics

a. Inertial Systems

Dynamics deals with the way bodies influence the motion of one another. Before pursuing this topic, it will be advantageous to consider an object isolated as much as possible from all other objects and effects. To measure the motion of the object requires the choice of a coordinate system to which its changing position can be referred. However, coordinate systems themselves may be in motion. For example, a coordinate system on earth rotates about the earth's axis and revolves around the sun. The question arises as to which of all imaginable coordinate systems is the most convenient to use. There is a special class of coordinate systems, called **inertial systems,** in which the laws of mechanics take an especially simple form. Inertial systems have the property that if one were to effectively isolate an object from all other objects and their effects, then the isolated object would have a constant velocity there. That is, an isolated body has no acceleration in an inertial system. Stars to an approximation are so distant from one another that they are isolated. Hence, any coordinate system at rest with respect to the stars or any system moving relative to this with constant velocity can be taken as an inertial system.

b. Mass

Loosely speaking, the **mass** of an object has to do with its resistance to being accelerated. A more precise definition of mass must include a description of the experimental process by which mass is to be measured. The operational definition presented here, originally due to Ernst Mach, involves the measurement of the accelerations produced when an object interacts with a reference object. Let us choose one particle as our reference particle R. We choose a second particle A and we isolate it and our reference particle as much as possible from all other effects while allowing the two to interact. We find that the two particles mutually influence the motion of each other, causing them to accelerate. As measured in an inertial coordinate system, the acceleration of the reference particle R due to its interaction with particle A is $\mathbf{a}_{RA}$, while the acceleration of particle A due to its interaction with reference particle R is $\mathbf{a}_{AR}$. Experiment shows that these accelerations may vary with the relative positions of the particles, with their velocities, and with time. However, it is found that the ratio $-a_{RA}/a_{AR}$ is a positive scalar that does not depend on position, velocity, or time as long as the particles are moving slowly compared with the speed of light. Now we remove particle A and repeat the same experiment with the reference particle R and another particle, B. The resulting accelerations of particle R and B are $\mathbf{a}_{RB}$ and $\mathbf{a}_{BR}$, respectively. Finally, we allow particles A and B to interact with each other, yielding accelerations a_{AB} and a_{BA} for particles A and B, respectively. It is found that

$$-a_{AB}/a_{BA} = \frac{-a_{RA}/a_{AR}}{-a_{RB}/a_{BR}} \tag{1}$$

which says that we can predict the ratio of accelerations of any two particles knowing only how each particle is accelerated relative to a reference particle.

We now define $-a_{RA}/a_{AR}$ to be the **mass** m_A of particle A relative to the reference. Similarly, $-a_{RB}/a_{BR}$ is the mass m_B of particle B. This definition coincides with the idea of an inverse relationship existing between the mass of a particle and its acceleration.

c. Force

In our everyday use of the word, a **force** is a push or a pull which sets a body in motion. More precisely, in an interaction of a particle of mass m with another particle, the quantity ma, where **a** is the acceleration of m due to the interaction, is called the force exerted on m by the other particle. Experiment shows that when several forces act upon a particle, each force acts independently of the others. Symbolically, we write

$$\mathbf{F} = m\mathbf{a} \tag{2}$$

where **F** stands for the vector sum of all forces acting on the object, m is its mass and **a** its acceleration. Equation 2 is a statement of Newton's second law of motion. Newton's first law is a special case of the second, namely, that when there is no force exerted on an object it moves with constant velocity in a straight line. This result has already been encountered in Example 2 and in the discussion of inertial systems.

Another important property of forces is revealed by an examination of Equation 1:

$$-a_{AB}/a_{BA} = \frac{-a_{RA}/a_{AR}}{-a_{RB}/a_{BR}}$$

Using the definition of mass, we have

$$-a_{AB}/a_{BA} = m_A/m_B$$

or

$$m_A\, a_{AB} = -m_B\, a_{BA} \tag{3}$$

This is a statement of Newton's third law: if two particles act on each other, the force exerted by the first on the second is equal in magnitude and opposite in direction to the force exerted by the second on the first.

A word about units: the mks system of units is currently the most favored among physical scientists. In this system length is measured in meters (m), mass in kilograms (kg), and time in seconds (sec). The force required to accelerate a 1-kg mass at the rate of 1 m/sec^2 is called the newton, having dimensions of mass × length/$time^2$.

d. Newton's Equations of Motion

An immense variety of mechanical problems can be treated using Newton's equations of motion as expressed by Equation 2:

$$\mathbf{F} = m\mathbf{a} = m \frac{d^2\mathbf{R}}{dt^2}$$

where **R** is the position vector of the particle. If a rectangular coordinate system is used, the component equations of motion are

$$F_x = m \frac{d^2 x}{dt^2}$$

$$F_y = m \frac{d^2 y}{dt^2}$$

$$F_z = m \frac{d^2 z}{dt^2}$$

where F_x, F_y, and F_z are the x- , y- , and z-components of the force **F** on the particle. It should be mentioned that Newton's equations of motion require modification when dealing with "nonclassical" situations such as systems moving at extremely high speeds or having extremely small dimensions.

To apply Equation 2 to a specific problem, it is necessary to know what to use for **F** on the left-hand side of the equation. Ultimately, one must guess an expression for the force which will lead to equations of motion that square with experimental observation. For example, based on the astronomical data of Tycho Brahe, Newton determined that the gravitational force existing between two objects is proportional to the product of their masses and to the inverse of the square of the distance between their centers. It is currently believed that all forces are manifestations of only four fundamental forces: gravitational, electrical, strong, and weak interactions.

Example 4. Simple Harmonic Motion

Consider a particle of mass m attached to a spring and moving along the x-axis, as illustrated in Figure 6. For small displacements x from the equilibrium position the force on the particle turns out to be approximately equal to −kx, where k is a positive constant called the spring constant. The minus sign signifies that the force is in a direction opposite to the displacement. Since the force and the acceleration of the particle are only along the x-axis, Equation 2 reduces to

$$-kx = m \frac{d^2 x}{dt^2} \qquad (4)$$

This differential equation states a relation between x and its second derivative with respect to time, viz., that the two are proportional to each other. Two functions of time which have this property are sin t and cos t since

$$\frac{d(\sin t)}{dt} = \cos t \qquad \frac{d(\cos t)}{dt} = -\sin t$$

$$\frac{d^2(\sin t)}{dt^2} = -\sin t \qquad \frac{d^2(\cos t)}{dt^2} = -\cos t$$

Although both of these functions are solutions of Equation 4, they are not the most general possible solution. A completely general solution has the form

$$x = A \sin(\omega t + \delta) \qquad (5)$$

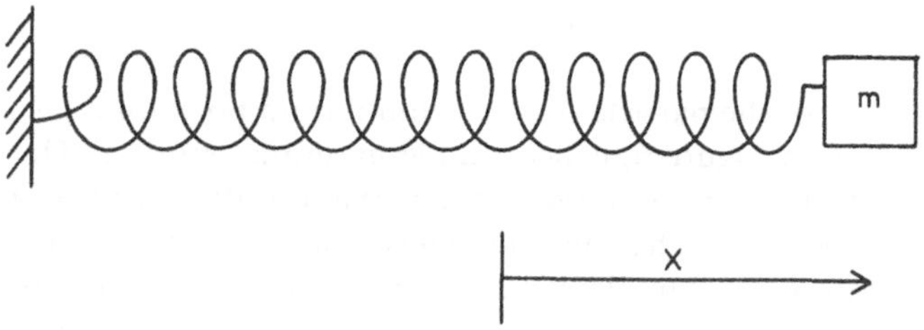

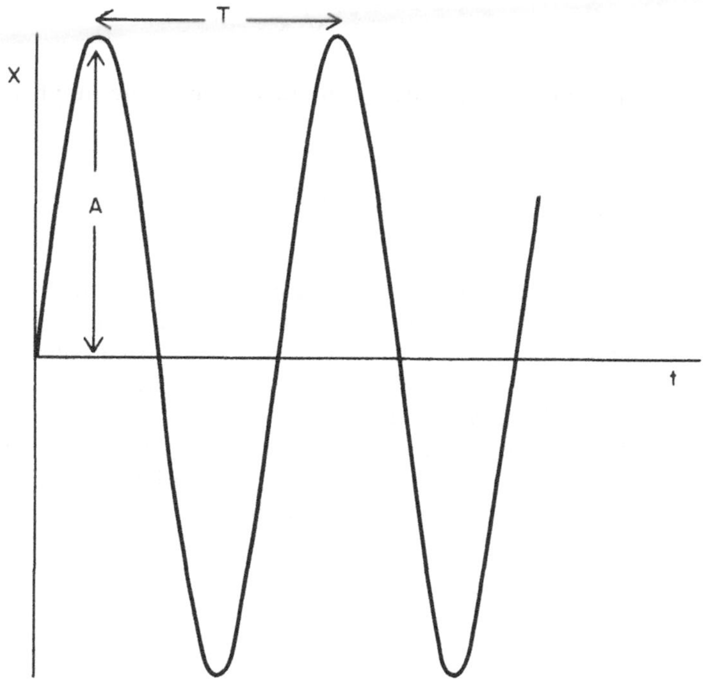

FIGURE 6. The simple harmonic motion of a mass attached to a spring. The sinusoidal displacement x of the mass from its equilibrium position as a function of time t is characterized by the amplitude A and period T. In this example the phase constant δ is zero since the displacement is zero at $t = 0$.

where A, ω, and δ are constants. Differentiating Equation 5 twice with respect to time gives

$$\frac{d^2 x}{dt^2} = -\omega^2 A \sin(\omega t + \delta)$$

and substituting into Equation 4, we have

$$-\omega^2 A \sin(\omega t + \delta) = -\frac{k}{m} A \sin(\omega t + \delta)$$

So the expression of Equation 5 satisfies Newton's equation of motion provided that

$$\omega^2 = k/m$$

So the displacement of the particle from its equilibrium position is a sinusoidal function of time as shown in Figure 6, called **simple harmonic motion.** The constant A is called the **amplitude** of the motion and is the maximum displacement of the particle from its equilibrium position. The constant ω has to do with the length of time it takes the particle to go through a complete cycle of motion, because if t in Equation 5 is increased by $2\pi/\omega$, the function $\sin(\omega t + \delta)$ exactly repeats itself. So the **period** T of the motion is $2\pi/\omega$:

$$T = \frac{2\pi}{\omega} = 2\pi\sqrt{\frac{m}{k}}$$

The inverse of the period is the **frequency** of oscillation f, the number of complete oscillations per unit time:

$$f = \frac{1}{T} = \frac{\omega}{2\pi}$$

The quantity δ is called the **phase constant** of the motion and has the effect of producing a time shift in the oscillatory motion of the particle without affecting its amplitude or period. For example, if $\delta = 0$, the particle is located at its equilibrium position ($x = 0$) at times $t = 0, T, 2T$, etc.; while if $\delta = \pi$, the particle is at its equilibrium position at times $t = T/2, 3T/2, 5T/2$, etc.

Besides the sine and cosine, there is another function which when differentiated twice yields the same function again and so is an appropriate solution of Newton's equation of motion for the simple harmonic oscillator. This function is e^{at} whose time derivatives are

$$\frac{d\,(e^{at})}{dt} = a\,e^{at}$$

and

$$\frac{d^2\,(e^{at})}{dt^2} = a^2\,e^{at}$$

Substituting into Equation 4, we have

$$-k\,e^{at} = m\,a^2\,e^{at}$$

so that

$$a = \sqrt{-\frac{k}{m}}$$

Since both k and m are positive, a must be an imaginary number

$$a = i\,\omega$$

where $i = \sqrt{-1}$ and $\omega = \sqrt{k/m}$. Now, imaginary exponents are related to the sine and cosine functions through the expression

$$e^{iy} = \cos y + i \sin y$$

as can be seen from an examination of the series expansions of e^{iy}, sin y, and cos y:

$$e^{iy} = 1 + \frac{iy}{1!} + \frac{(iy)^2}{2!} + \frac{(iy)^3}{3!} + \cdots$$

$$\sin y = y - \frac{y^3}{3!} + \frac{y^5}{5!} - \frac{y^7}{7!} + \cdots$$

$$\cos y = 1 - \frac{y^2}{2!} + \frac{y^4}{4!} - \frac{y^6}{6!} + \cdots$$

We take as the general solution of Newton's equation of motion for the mass on a spring the complex variable x':

$$x' = A\, e^{i(\omega t + \delta)} \qquad (6)$$

where, as before, A is a real number representing the amplitude of motion, ω is related to the frequency f by $f = \omega/2\pi$, and δ is the phase constant. We may rewrite Equation 6 as

$$x' = A\, e^{i\omega t}\, e^{i\delta}$$

$$= A'\, e^{i\omega t} \qquad (7)$$

where the complex amplitude A' contains both the amplitude and phase information:

$$A' = A\, e^{i\delta}$$

Using an exponential function of time such as Equation 7 rather than a trigonometric expression has many mathematical advantages since the operations of multiplication, division, and differentiation are less cumbersome. Of course, some convention is needed whereby a correspondence can be made between the complex variables and the real physical quantities involved, such as displacement and velocity. What is done is simply to relate the physical displacement at time t to the real part of x', etc. A word of caution is that the square of a physical quantity is found by squaring the real part of the corresponding complex variable rather than by taking the real part of the square of the complex variable.

Example 5. Kepler's Third Law of Planetary Motion

Consider a planet moving around the sun in a circular orbit, which is approximately the case for the earth. According to Newton's law of gravitation, the magnitude of the gravitational force exerted by the mass of the sun m_s on the mass of the planet m_p is

$$F = \frac{G\, m_s\, m_p}{R^2} \qquad (8)$$

where G is the constant of universal gravitation and R is the distance from the sun to the planet. We also know from Example 3 that for a particle to move in a circular path with constant velocity v, a force is required to be directed toward the center of the circle of magnitude $m_p v^2/R$. Hence,

$$\frac{G\, m_s m_p}{R^2} = \frac{m_p\, v^2}{R}$$

or

$$v^2 = \frac{G\, m_s}{R}$$

Now, the time it takes the planet to make a complete trip around the sun is called the period T. Since the circumference of the circle is $2\,\pi\,R$,

$$T = \frac{2\pi R}{v} = \frac{2\pi R}{\sqrt{\dfrac{G\, m_s}{R}}}$$

or

$$T^2 = \frac{4\pi^2 R^3}{G\, m_s}$$

Hence, the square of the period of any planet about the sun is proportional to the cube of the distance of the planet from the sun. This is known as Kepler's third law of planetary motion.

e. Linear Momentum

Since the concept of force is so fundamental to mechanics, it is natural to inquire about its integrated effect on a particle or system of particles over time. Integrating the force on a particle from time t_1 to time t_2, we have

$$\int_{t_1}^{t_2} \mathbf{F}\, dt = \int_{t_1}^{t_2} m\, \mathbf{a}\, dt$$

$$= m \int_{t_1}^{t_2} \frac{d\mathbf{v}}{dt}\, dt$$

$$= m\,(\mathbf{v}_2 - \mathbf{v}_{-1}) \qquad (9)$$

where $\mathbf{v}_1$ and $\mathbf{v}_2$ are the velocities of the particle at t_1 and t_2, respectively. The quantity mv, the product of the mass of an object and its velocity, is called the **linear momentum** of the object. Evidently, the time integral of the force on a particle is equal to the change in linear momentum of the particle. Equivalently, the rate of change of the momentum of a particle is equal to the force on the particle:

$$\mathbf{F} = \frac{d}{dt}\,(m\,\mathbf{v}) \qquad (10)$$

To understand the usefulness of the idea of momentum, let us consider two objects isolated from the rest of the universe that are exerting a force on one another. From Equation 3, the force $\mathbf{F}_1$ acting on object 1 is accompanied by an equal and opposite force $\mathbf{F}_2$ acting on object 2. Suppose object 1 has mass m_1 and velocity **u** while object 2 has mass m_2 and velocity **v**. From Equation 9 the time integral of the sum of the forces on the objects between times t_A and t_B is equal to the sum of the changes in the momenta of the objects. But the sum of the forces is zero so that

$$\int_{t_A}^{t_B} (\mathbf{F}_1 + \mathbf{F}_2)\, dt = 0 = m_1 (\mathbf{u}_A - \mathbf{u}_B) + m_2 (\mathbf{v}_A - \mathbf{v}_B) \qquad (11)$$

Since the times t_A and t_B are arbitrary, Equation 11 states that the change in the total momentum of the two objects is always zero.

Extending this line of argument to a collection of more than two objects, we can always sort out the forces that the objects exert on one another into action and reaction pairs. Hence we arrive at the **principle of the conservation of linear momentum**: as long as no external forces act on a system of objects, the total linear momentum of the system, defined as the vector sum of the individual momenta, remains constant. An external force is one that acts on one or more objects in the system but originates outside the system. Even if there is an external force acting on the system, there may be directions along which the vector components of the force are zero. In that case the corresponding vector components of the total momentum do not change with time. The principle of the conservation of momentum is useful in those situations where the nature of the internal forces acting between particles is not known but the momenta of the particles are.

Example 6. Completely Inelastic Collision

A block of wood of mass M lies at rest on a horizontal surface. It is struck by a bullet of mass m traveling horizontally with velocity v. The bullet becomes embedded in the block. Suppose it is possible to ignore any frictional effects so that there are no external forces with a horizontal component acting on the objects. Then the principle of the conservation of momentum states that the initial and final momenta are the same:

$$mv = (M + m)\, u$$

where u is the final velocity of the bullet-block combination. Hence,

$$u = \frac{m}{M + m}\, v$$

Here, the principle of the conservation of momentum has allowed the motion of a system of objects to be determined even though the forces acting between them are unknown.

f. Torque

It is of interest to consider the force on an object that is effective in causing the object to rotate about an axis. For example, the force most effective in causing a door to swing on its hinges is one which is applied perpendicular to the door and as far as possible from the hinges. That which makes something rotate about an axis is called a

torque. We define torque in the following way. If a force **F** acts on a particle whose position with respect to some point is specified by the position vector **R**, then the torque **N** on the particle about the point is the cross product

$$\mathbf{N} = \mathbf{R} \times \mathbf{F}$$

We recall (Section II.B.3) that the magnitude of **N** is

$$N = R\,F \sin\theta$$

where θ is the angle between **R** and **F**. We note that the torque is a maximum when (1) **F** is perpendicular to **R** and (2) the lever arm R and the force F are as large as possible. The direction of **N** is perpendicular to both **R** and **F** so that if one was to rotate **R** into **F** with the fingers of the right hand, then the thumb would point in the direction of **N**.

g. Angular Momentum

Mathematically, the **moment** of a quantity is found by weighting the quantity according to how far away it is from some point or line. Torque, then, is the moment of the force. Another important quantity is the moment of the linear momentum, called the **angular momentum.** If the position of a particle relative to some point is specified by the vector **R** and if the particle has a linear momentum **p** = m**v**, then the angular momentum **L** of the particle about the point is defined as

$$\mathbf{L} = \mathbf{R} \times \mathbf{p}$$

Hence, the magnitude of the angular momentum is

$$L = R\,p \sin\theta$$

where θ is the angle between **R** and **p**. The direction **L** is perpendicular to **R** and **p** as specified by the right-hand rule.

Just as the force on a particle is related to its change in linear momentum, a similar relation exists between torque and angular momentum. To show this relation, we find the time derivative of the angular momentum, using the fact that the derivative of a cross product is taken in the same way as the derivative of an ordinary product:

$$\begin{aligned}\frac{d\mathbf{L}}{dt} &= \frac{d}{dt}(\mathbf{R} \times \mathbf{p}) \\ &= \frac{d\mathbf{R}}{dt} \times \mathbf{p} + \mathbf{R} \times \frac{d\mathbf{p}}{dt} \\ &= \mathbf{v} \times m\mathbf{v} + \mathbf{R} \times m\frac{d\mathbf{v}}{dt}\end{aligned}$$

The term **v** × m**v** is zero since the cross product of two parallel vectors is zero. The term **R** × m (d**v**/dt) is just **R** × **F**, which is the torque on the particle. Hence,

$$\frac{d\mathbf{L}}{dt} = \mathbf{N} \qquad (12)$$

in analogy with Equation 10:

$$\frac{dp}{dt} = F$$

Equation 12 may be extended to a system of particles, in which case **L** and **N** represent the vector sum of the angular momenta and torques, respectively, of all the particles.

For the special case in which the net torque on the system of particles is zero, the total angular momentum of the system remains constant. There is an especially important circumstance in which this happens. Suppose every pair of particles in the system exerts forces on each other which by Newton's third law are equal and opposite but which also happen to point along the line joining each pair. Gravitational and electrostatic forces are examples of these kinds of "central forces". In this case it is straightforward to show that the sum of the torques due to these internal forces is zero. This leads to the **principle of the conservation of angular momentum,** viz., for internal forces having the properties described above, if there is no external torque acting on a system of particles, the total angular momentum of the system remains constant.

Example 7. Kepler's Second Law of Planetary Motion

In the motion of the earth about the sun, the much greater mass of the sun causes it to remain nearly stationary while the earth revolves around it. Let us consider these two bodies as the system of interest. The forces exerted on the earth and sun by other planets, moons, stars, etc., are small enough so that to a good approximation there are no external torques acting on the system. Hence, the angular momentum of the system remains constant.

If the mass of the earth is m, its distance from the sun R, and its speed v, then the angular momentum of the earth about the sun has the magnitude

$$L = R\,m\,v \sin\theta$$

where θ is the angle between **R** and **v** as shown in Figure 7. The origin of the coordinate system has been taken to be at the sun. In moving a small distance $v\Delta t$ in the time interval Δt, a certain area will be swept out by the position vector of the earth. This area is approximately a triangle. From trigonometry it may be recalled that the area of a triangle is ½ AB sin θ, where θ is the angle between the sides of length A and B. For our triangle one side is of length R, the other of length $v\Delta t$, so that the small area (Δ area) swept out by the position vector in time Δt is

$$\Delta\,\text{area} = \tfrac{1}{2} R\,v\,\Delta t \sin\theta$$

or

$$\frac{\Delta\,\text{area}}{\Delta t} = \tfrac{1}{2} R\,v \sin\theta$$

$$= \tfrac{1}{2}\frac{L}{m}$$

Since L and m are constants, we have derived Kepler's second law of planetary motion, which states that a line joining a planet to the sun sweeps out equal areas in equal times.

h. Energy

As we have seen, the integral of force over time is the very useful quantity called

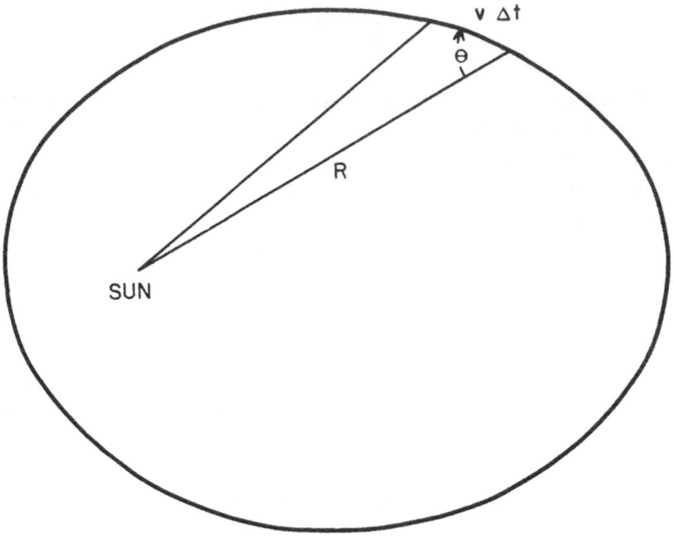

FIGURE 7. Kepler's second law of planetary motion in which the line joining a planet to the sun sweeps out equal areas in equal times.

momentum. We now wish to investigate the integral of force along a path in space that leads to the equally important concept of **energy**. Suppose a particle of mass m experiences a force **F** in moving through a small displacement d**R**. We consider the summation of the dot products **F** · d**R** along a particular path of the particle from point P_1 to point P_2. This quantity is called the **work** done on the particle by the force:

$$\text{Work} = \int_{P_1}^{P_2} \mathbf{F} \cdot d\mathbf{R} \tag{13}$$

Work is a scalar quantity that is a maximum when the force is in the direction in which the particle is moving, is zero when the force is perpendicular to the path of the particle, and is negative when the force is in a direction opposite to that in which the particle is moving. For example, the work done on the earth by the gravitational attraction of the sun is zero since the path of the earth approximates a circle and the force of attraction is always at right angles to the direction of motion.

Substituting ma = mdv/dt for **F** in Equation 13 and vdt for d**R**, we have

$$\int_{P_1}^{P_2} \mathbf{F} \cdot d\mathbf{R} = \int_{P_1}^{P_2} \left(m \frac{d\mathbf{v}}{dt}\right) \cdot (\mathbf{v}\, dt)$$

$$= m \int_{P_1}^{P_2} \mathbf{v} \cdot d\mathbf{v}$$

$$= \tfrac{1}{2} m v_2^2 - \tfrac{1}{2} m v_1^2 \tag{14}$$

where v_1 and v_2 are the speeds of the particle at points P_1 and P_2, respectively. Now, the **kinetic energy** of a particle is defined to be the quantity ½mv^2:

$$\text{Kinetic energy} = \tfrac{1}{2} mv^2$$

Equation 14, therefore, says that the line integral of the force over the path taken by the particle, i.e., the work done on the particle, is equal to the change in kinetic energy of the particle. Hence, a body possesses kinetic energy (or energy of motion) because work has been done on the body, i.e., a force has been exerted on the body over a certain distance that has speeded it up or slowed it down. Conversely, because of Newton's third law, the work done on the body by the force is the negative of the work done by the body on whatever produced the force. So the kinetic energy that a body possesses can be thought of as the work that the body can do on something else by virtue of its motion.

Before defining potential energy, it will be useful to review in a simplified way some of the basic ideas of the differential calculus of functions of several variables. Suppose the value of a function U depends on the values of the variables x, y, and z. The **partial derivative** of U with respect to one of the variables, say x, is found by holding the values of the other variables constant and taking the derivative of U with respect to x as in ordinary calculus. Thus, if $U = x^3 + xyz$, then the partial derivative of U with respect to x (written $\partial U/\partial x$) is

$$\frac{\partial U}{\partial x} = 3x^2 + yz$$

The **total differential** of U, dU, relates what small change will result in U when the variables are changed by the small increments dx, dy, and dz. This relationship is

$$dU = \frac{\partial U}{\partial x} dx + \frac{\partial U}{\partial y} dy + \frac{\partial U}{\partial z} dz \tag{15}$$

Since $\int dU = U$, we have for the integration from point P_1 to point P_2:

$$\int_{P_1}^{P_2} dU = \int_{P_1}^{P_2} \left(\frac{\partial U}{\partial x} dx + \frac{\partial U}{\partial y} dy + \frac{\partial U}{\partial z} dz \right) = U_2 - U_1 \tag{16}$$

where U_2 and U_1 are the values of U at points P_2 and P_1, respectively.

Now, returning to the line integral of the force $\int \mathbf{F} \cdot d\mathbf{R}$, we write **F** and d**R** in terms of their x-, y-, and z-components:

$$\mathbf{F} = F_x \mathbf{i} + F_y \mathbf{j} + F_z \mathbf{k}$$

and

$$d\mathbf{R} = dx\,\mathbf{i} + dy\,\mathbf{j} + dz\,\mathbf{k}$$

Taking the dot product, we have

$$\mathbf{F} \cdot d\mathbf{R} = F_x\,dx + F_y\,dy + F_z\,dz$$

so that

$$\int_{P_1}^{P_2} \mathbf{F} \cdot d\mathbf{R} = \int_{P_1}^{P_2} (F_x\,dx + F_y\,dy + F_z\,dz) \tag{17}$$

We now compare Equation 16 with Equation 17. What if it were possible to find a function U such that the total differential of U were equal to $F_x\,dx + F_y\,dy + F_z\,dz$? If such a function U existed, then $\int \mathbf{F} \cdot d\mathbf{R}$ by Equation 16 would depend only on the value of U at the beginning point P_1 of the path and the end point P_2. Actually, it is customary to look for the negative of this function such that

$$-dU = F_x\,dx + F_y\,dy + F_z\,dz$$

Then from Equation 15

$$F_x = -\frac{\partial U}{\partial x}$$

$$F_y = -\frac{\partial U}{\partial y}$$

$$F_z = -\frac{\partial U}{\partial z} \tag{18}$$

The function U, when it exists, is called the **potential energy** of the particle and can be found with the help of Equation 18. Equation 16 then becomes

$$\int_{P_1}^{P_2} \mathbf{F} \cdot d\mathbf{R} = U_1 - U_2 \tag{19}$$

where U_1 and U_2 are the values of the potential energy at points P_1 and P_2, respectively. Equation 19 says that the work done on a particle by a force in traveling any route between points P_1 and P_2 is the difference in potential energy between the two points.

We note that an arbitrary constant can be added to U and Equation 18 will still be satisfied. So the value of the potential energy is arbitrary to within an additive constant, it being the change in potential energy that has physical significance. There is a similar arbitrariness in the value of the kinetic energy, since the measured speed of an object depends on the motion of the observer.

In contrast to kinetic energy, potential energy depends on the position of the particle rather than on its speed. The change in potential energy of an object is equal to the work required to move the particle from one position to another. Conversely, the amount of work that the body can do on something else by virtue of a change in its position is equal to its change in potential energy.

In general, when work is done on an object, both its potential and kinetic energies change and the change in each is equal to the work done on the object:

$$\int_{P_1}^{P_2} \mathbf{F} \cdot d\mathbf{R} = U_1 - U_2 = \tfrac{1}{2}mv_2^2 - \tfrac{1}{2}mv_1^2$$

or

$$U_1 + \tfrac{1}{2}mv_1^2 = U_2 + \tfrac{1}{2}mv_2^2 \tag{20}$$

The sum of the potential and kinetic energy is called the **total energy** E of the particle.

$$E = \frac{1}{2}mv^2 + U$$

Equation 20 states that if the function U exists, the total energy of the particle does not change from point to point in its motion. Any increase in the kinetic energy of the particle, for example, must be accompanied by a corresponding decrease in the potential energy. This is the **principle of the conservation of energy.** It can easily be extended to a system of particles. Like the conservation principles for linear and angular momentum, the principle of conservation of energy is useful because it delineates a quantity that, under the restrictions mentioned above, remains constant regardless of the complexities of the interactions among the particles constituting the system. It, along with the other conservation principles may be used instead of, or in addition to, Newton's laws of motion to treat the various problems of particle dynamics.

Under what circumstances does a function U exist whose total differential is $-\mathbf{F}\cdot d\mathbf{R}$? As can be seen from Equation 19, it is necessary that the work done by the force along a path between two points depend only on the location of these points and not on the path followed. A simple extension of this idea is that if the beginning and end point of the path are the same so that the particle has made a round trip, the net work done by the force is zero. Forces for which the work done in moving a body between two given points is independent of the path are said to be **conservative.** For these forces a potential energy function exists; otherwise the forces are **nonconservative.**

Frictional forces are nonconservative since, in always acting to oppose the motion of an object, the work that they do depends on the path taken. In general, frictional forces do negative work and the longer the path length, the greater is this negative quantity. A closer inspection of friction, however, reveals that in the process of opposing the motion of the object, the molecules of the object and of the surrounding medium acquire increased kinetic energy so that in a complete accounting, energy is exactly conserved. In fact, whenever the effects of forces are studied in great enough detail, it always seems to be true that the total energy is exactly conserved. This is a consequence of the fact that, as far as is known, the fundamental forces of nature are all conservative.

Energy has dimensions of mass × length²/time². The unit of energy in the mks system is the joule (J).

Example 8. Energy of a Simple Harmonic Oscillator

The example of a mass m attached to a spring was discussed in Section III.A.2.d. The force on the mass is $-kx\mathbf{i}$, where k is the spring constant, x is the displacement of the mass from the equilibrium position, and **i** is the unit vector in the x-direction. We found that the x-coordinate as a function of time was given by the sinusoidal function

$$x = A \sin(\omega t + \delta)$$

where A is the amplitude of the motion, $\omega = \sqrt{k/m}$, and δ is the phase. Now, the force $-kx\mathbf{i}$ is conservative since there exists a potential energy function

$$U = \frac{1}{2}kx^2$$

having the desired property that

$$\frac{\partial U}{\partial x} = -F_x$$

$$\frac{\partial U}{\partial y} = F_y = 0$$

$$\frac{\partial U}{\partial z} = F_z = 0$$

Substituting for x^2, we have for the potential energy

$$U = \tfrac{1}{2}kA^2 \sin^2 (\omega t + \delta)$$

The kinetic energy at any time is

$$\begin{aligned} \tfrac{1}{2}m v^2 &= \tfrac{1}{2}m \left(\frac{dx}{dt}\right)^2 \\ &= \tfrac{1}{2}m \,[A \,\omega \cos (\omega t + \delta)]^2 \\ &= \tfrac{1}{2}m\, A^2 \omega^2 \cos^2 (\omega t + \delta) \\ &= \tfrac{1}{2}k\, A^2 \cos^2 (\omega t + \delta) \end{aligned}$$

The total energy E of the system is the sum of the potential and kinetic energies:

$$\begin{aligned} E &= \tfrac{1}{2}kA^2 \cos^2 (\omega t + \delta) + \tfrac{1}{2}kA^2 \sin^2 (\omega t + \delta) \\ &= \tfrac{1}{2}kA^2 \end{aligned}$$

So the total energy of the mass on the spring is indeed constant, being proportional to the spring constant k and the square of the amplitude of the motion A^2.

Example 9. Energy of the Hydrogen Atom

In the hydrogen atom an electron orbits about a much more massive proton. As will be discussed in Section III.B.1, the attractive force existing between electron and proton is the Coulomb force:

$$\mathbf{F} = -\frac{k}{R^2}\,\mathbf{r}$$

where k is a constant that depends on the electric charge of the electron and proton, R is the distance between the two particles, and r is a unit vector pointing from proton to electron. The potential energy U of the electron is

$$U = -\frac{k}{R} \qquad (21)$$

since in that case $\partial U/\partial R = -F_R$. The kinetic energy T of the electron is

$$T = \tfrac{1}{2}\, mv^2$$

where m is the mass of the electron and v its velocity. If the orbit of the electron is circular, from Example 3 we have

$$F = \frac{k}{R^2} = \frac{mv^2}{R} \tag{22}$$

or

$$T = \tfrac{1}{2}mv^2 = \tfrac{1}{2}\frac{k}{R}$$

So the total energy E of the electron is

$$\begin{aligned} E &= T + U \\ &= \frac{k}{2R} - \frac{k}{R} \\ &= -\frac{k}{2R} \end{aligned}$$

As expected, the total energy E is a constant. What may be unexpected is that E is negative. In Equation 21 we chose not to add an arbitrary constant to the potential energy U so that U is always negative, tending toward zero as R becomes infinitely large. The kinetic energy T, which must always be positive, in this case is only half as large as U. Under these circumstances a negative value for E indicates that the electron is **bound** to the proton, i.e., it remains a finite distance away. The **binding energy** of the electron is defined as the positive energy required to separate the electron completely from the proton.

Surprisingly, measurements of the energy of the electron in the hydrogen atom show that only certain special values are possible for E. Bohr was the first to demonstrate that the angular momentum of the electron is also restricted to special values, since if the angular momentum L is an integral multiple of a constant $\hbar$, then the correct values for the energy emerge. That is, if

$$L = mvR = n\hbar \tag{23}$$

where $n = 1, 2, 3, \cdots$, and Planck's constant $\hbar = 1.05 \times 10^{-34}$ J/sec, then after eliminating v from Equations 22 and 23 to solve for R, we have

$$E = -\frac{mk^2}{2\hbar^2}\frac{1}{n^2}$$

which is confirmed by experiment. The restriction of the energy and angular momentum of very small systems to special values is the subject of the **quantum theory** (Section IV.B).

B. Electricity and Magnetism

1. Electric Charges and Forces

We shall now consider the forces that arise from certain electric and magnetic phenomena. If one rubs two glass rods with silk and two rubber rods with fur, the glass rods will repel each other as will the rubber rods. But a glass rod and a rubber rod will attract each other. We explain these facts by saying that rubbing any of these rods gives it an **electric charge,** of which there are at least two kinds. Furthermore, like charges repel and unlike charges attract. Benjamin Franklin termed the charge on the glass "positive" and the charge on the rubber "negative" and these designations have remained. Today we know that there are only two kinds of electric charges and that as far as the basic constituents of matter are concerned electrons are negatively charged, protons are positively charged, and neutrons have no charge. Normally, matter contains equal numbers of protons and electrons and so is electrically neutral. But rubbing two objects together can result in transferring electrons from one object to another, producing a net charge on each.

The magnitude of the force between two stationary particles having charges q_1 and q_2 was shown by Coulomb to be

$$F = k \frac{q_1 q_2}{R^2}$$

where k is a constant and R is the distance between the particles. The direction of the electrostatic **Coulomb force** lies along the line joining the two particles, acting so as to attract them if the product q_1q_2 is negative and to repel them if q_1q_2 is positive. The similarity between the mathematical form of Coulomb's law and Newton's law of gravitation (Equation 8) is apparent.

The value of the proportionality constant k depends on the units chosen. In the mks system the charge is in coulombs (C), the force in newtons (N), the distance in meters (m) and

$$k = \frac{1}{4\pi\epsilon_o}$$

where $\varepsilon_o = 8.85 \times 10^{-12}\ C^2/N/m^2$. The factor of 4π is included to make other important formulas simpler.

2. Electric Field

Two charged particles in space exert a force on one another as described by Coulomb's law. Now, instead of imagining that each particle acts directly on the other, an alternative way of viewing the situation is that the particles produce in the surrounding space a condition called an **electric field.** It is this field in turn which exerts a force on the particles. For stationary charges little is gained by this formulation. However, if one of the particles is moving with respect to the other, electromagnetic theory predicts that the second particle cannot instantaneously learn about the movement of the first particle. Instead, the information travels through space at the speed of light. It is convenient to view this transfer of information as a disturbance propagating through the electric field.

We define the electric field E at any point to be the ratio of the electric force on a test charge placed there to the value of the test charge:

$$E = \frac{F\,(\text{on } q)}{q} \qquad (24)$$

where q is small enough so that it does not significantly disturb the primary charges that are responsible for the field. For example, the electric field at some point a distance R from a stationary charge q is

$$\mathbf{E} = \frac{1}{4\pi\epsilon_o} \frac{q}{R^2} \mathbf{r}$$

where r is a unit vector pointing from the charge to the point.

The electric field unit is the newton/coulomb.

Example 10. The Electric Dipole

Certain combinations of charges are of special interest. One of these is the **electric dipole** which consists of two equal but opposite charges +q and −q, placed a distance D apart. To calculate the electric field **E** at a point P equidistant from the two charges we simply add the electric fields due to each charge:

$$\mathbf{E} = \mathbf{E}_1 + \mathbf{E}_2$$

The magnitude of each field is given by

$$E_1 = E_2 = \frac{1}{4\pi\epsilon_o} \frac{q}{R^2}$$

where R is the distance of the point P from each charge. As shown in Figure 8, the components of vectors $\mathbf{E}_1$ and $\mathbf{E}_2$ taken along the line joining P with the midpoint between the charges are equal in magnitude and opposite in direction and therefore cancel. The total field **E** then is in a direction perpendicular to this line and has magnitude

$$E = 2\,E_1 \cos\theta$$

Now,

$$\cos\theta = \frac{D}{2R}$$

so that

$$E = \frac{2}{4\pi\epsilon_o} \frac{q}{R^2} \frac{D}{2R}$$

$$= \frac{1}{4\pi\epsilon_o} \frac{p}{R^3}$$

where p = qD and is called the **electric dipole moment.** Thus the electric field due to an electric dipole is proportional to $1/R^3$.

An **electric quadrupole** consists of two electric dipoles. Its field is inversely proportional to the fourth power of the distance. In a similar way the electric octupole and higher orders of charge distribution can be defined. The utility of electric multipoles arises from the fact that an arbitrary charge distribution can, at large distances, be represented by a sum of monopole, dipole, quadrupole, etc., configurations.

3. Magnetic Field

The discussion to this point has assumed that all electric charges are stationary. In the event that there are moving charges present in the region then additional forces

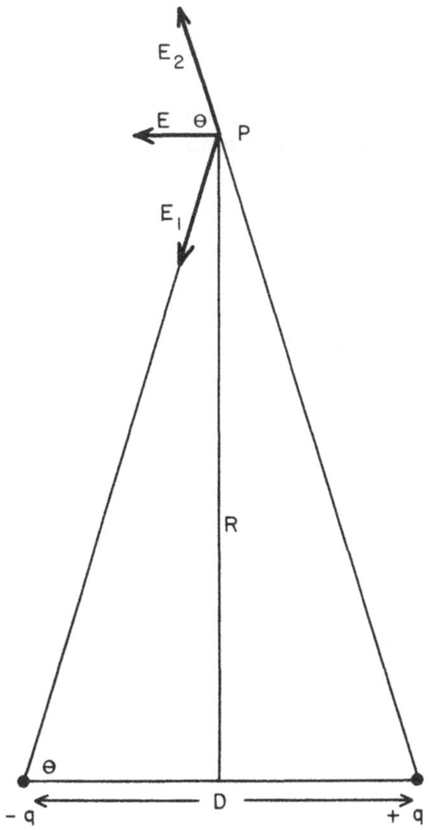

FIGURE 8. Electric field **E** at a point P due to an electric dipole. The distance R of the point from the dipole is much greater than the separation D of the charges.

arise and Coulomb's law is no longer valid. Experiment shows that a test charge moving through a region where other moving charges are present will experience one component of force that is perpendicular to its direction of motion. This force is termed **magnetic**. Magnetic forces, in addition, are proportional to the speed of the particle and to its charge. In analogy with the electric field, we define a **magnetic field, B,** as a condition of space which gives rise to the magnetic force $\mathbf{F}_M$ on a moving test charge. However, it does not prove advantageous to take **B** and $\mathbf{F}_M$ in the same direction since the magnetic field would then depend on the arbitrary direction of motion of the test particle. Instead, another fact is used to define the direction of the magnetic field **B**. Experiment shows that when magnetic forces are present there exists at each point in space one direction in which, if the test particle were projected along it, the particle would experience no velocity-dependent force. This direction is taken to be the direction of **B**.

The properties of the magnetic force acting on a small moving charge are such that it is (1) proportional to the charge q, (2) perpendicular to the velocity **v**, (3) proportional to the magnitude of the velocity, and (4) zero when **v** is along one certain direction. These properties can be compactly expressed by the vector cross product:

$$\mathbf{F}_M = q\,\mathbf{v} \times \mathbf{B}$$

Now, the fact that moving charges may be present does not alter the way in which the electric field **E** is defined. One continues to use a stationary test charge, and relates the force on it to the electric field according to Equation 24:

$$\mathbf{E} = \frac{\mathbf{F}_E}{q}$$

Hence, the total force **F** on a particle having charge q in an electric field **E** and magnetic field **B** is

$$\mathbf{F} = q\,\mathbf{E} + q\,\mathbf{v} \times \mathbf{B}$$

an expression known as the **Lorentz law.**

Not only do moving charges experience magnetic forces but it can be shown experimentally that moving charges produce magnetic forces as well. Suppose a conductor carrying an assembly of moving charges (a current) is divided up conceptually into many current elements of length dL. Then the contribution d**B** of a current element to the total field **B** at a point is given by the **Biot-Savart law:**

$$d\,\mathbf{B} = \frac{\mu_o i}{4\pi}\,\frac{d\mathbf{L} \times \mathbf{r}}{R^2} \qquad (25)$$

where the conductor is carrying current i, d**L** is in a direction tangent to the current, **r** is a unit vector pointing from the current element to the point of observation, R is the distance from the current element to the point, and μ_o is called the permeability constant. The field due to the entire conductor is then found by integrating Equation 25 over all current elements. In the mks system of units B is in webers/meter² (Wb/m²), current is in amperes (C/sec), and the permeability constant has the value $\mu_o = 4\pi \times 10^{-7}$ Wb/C/sec/m.

Example 11. The Magnetic Dipole

Let us compute **B** at a point P on the axis of a circular loop of diameter D carrying a current i. As shown in Figure 9, d$\mathbf{B}_1$ is the contribution due to a current element at the top of the loop and d$\mathbf{B}_2$ the contribution from an element at the bottom. Then the components of d$\mathbf{B}_1$ and d$\mathbf{B}_2$ perpendicular to the axis of the loop are equal in magnitude and opposite in direction and cancel. Only components along the axis contribute. This is true for any pair of opposed current elements, the magnitude of the contribution from each element being

$$dB = \frac{\mu_o i}{4}\,\frac{dL}{R^2}\cos\theta$$

But

$$\cos\theta = \frac{D}{2R}$$

so

$$dB = \frac{\mu_o i}{4\pi}\,\frac{D\,dL}{2R^3}$$

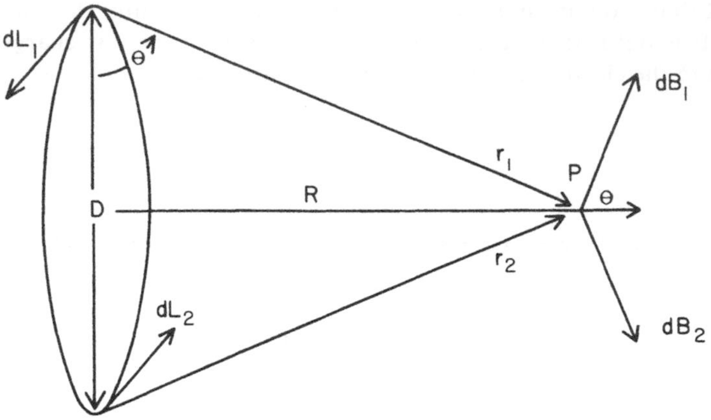

FIGURE 9. Magnetic field B at a point P due to a circular current loop. The distance R of the point from the current loop is much greater that the diameter D of the loop.

Integrating around the loop,

$$B = \frac{\mu_o i}{4\pi} \frac{D}{2R^3} \int dL$$

$$= \frac{\mu_o i}{4\pi} \frac{D}{2R^3} \pi D$$

Now, the area A of the loop is $\pi D^2/4$ and calling $\mu = Ai$, we have

$$B = \frac{\mu_o}{2\pi} \frac{\mu}{R^3}$$

This relation is reminiscent of the electric field due to an electric dipole (Example 10) which is also inversely proportional to R^3. Because of this and other similarities to the electric dipole, a plane current loop is said to constitute a **magnetic dipole** with **magnetic dipole moment** μ. By this definition a charged particle moving in a closed path in a plane also behaves like a magnetic dipole. A charged body, such as a sphere which simply spins about an axis, can be considered to be the summation of many plane current loops perpendicular to the axis. Since each loop is a magnetic dipole, a spinning charge possesses a magnetic dipole moment. Anticipating discussions of the atom, we see how it is plausible that an electron or proton moving in orbit or spinning about its axis can have magnetic dipole moments associated with these motions.

4. Electromagnetic Radiation

As discussed in Section III.B.1, a stationary electric charge q produces an electric field at a point a distance R away, of

$$\mathbf{E} = \frac{1}{4\pi\epsilon_o} \frac{q}{R^2} \mathbf{r}$$

E is spherically symmetric and, being inversely proportional to the square of the distance from the charge, rapidly diminishes in effect. No magnetic field is observed by a person who is stationary relative to the charge:

$$\mathbf{B} = 0$$

If the charge moves at constant velocity with respect to an observer, the electric field becomes more intense in the direction perpendicular to its velocity. In addition, a magnetic field is produced which exerts a force on other charged particles moving with respect to the observer. Both the electric and magnetic fields due to a charge moving with constant velocity are inversely proportional to the square of the distance from the charge to the observer.

When the charge is accelerating, something of great practical significance happens. Additional electric and magnetic field components are produced which have a 1/R variation with distance so that they can affect charges far away. When the velocity of the charge is small, these fields have the form:

$$\mathbf{E} = \frac{q}{4\pi\epsilon_o c^2} \frac{\mathbf{r}' \times (\mathbf{r}' \times \mathbf{a}')}{R'}$$

$$\mathbf{B} = \frac{q}{4\pi\epsilon_o c^3} \frac{\mathbf{a}' \times \mathbf{r}'}{R'} \qquad (26)$$

where r′ is a unit vector pointing from the charge to the point of observation, R′ is the distance from the charge, **a′** is the acceleration of the charge, and c is the speed of light ($c = 3.00 \times 10^8$ m/sec). The significance of the primes on r′, R′, and a′ is that, since information cannot travel faster than the speed of light, a particular motion of the charge at a particular time produces fields at a distance R at time R/c later. So to compute E at time t we evaluate the primed quantities at the earlier (or **retarded**) time (t − R/c). It is seen that **E** and **B**, due to an accelerating charge, are mutually perpendicular. For example, for a charge accelerating along the + z-axis, at a point on the + x-axis **E** is in the −z direction while **B** is in the + y direction.

Now, electric and magnetic fields have energy effects on electric charges since they can perform work on the charges. These energy effects happen to be proportional to the square of the field. We see that this is true in the simple example of a particle with an electric charge that is attached to a spring. In the presence of an electric field, a force is exerted on the charge causing it to be displaced a distance A which is proportional to the field. If the field is removed so that the particle oscillates, the displacement and velocity of the particle at a particular point in its cycle are proportional to A and hence to the initial field. The total energy of the particle, which as we have seen in Example 8 depends on the square of the displacement and the square of the velocity, is proportional to A^2 and hence to the square of the field.

We may imagine the energy effects that electromagnetic fields can deliver to reside in the fields themselves. That is, if an electric or magnetic field exists at some point in space, we think of stored energy existing there in an amount per unit volume that is proportional to the square of the field. The exact relationships are

$$u_E = \frac{\epsilon_o}{2} E^2$$

$$u_B = \frac{1}{2\mu_o} B^2$$

where u_E and u_B are the energies per unit volume stored in the electric and magnetic fields, respectively.

As a charge begins to accelerate, the fields that this motion creates are not felt immediately at some distant point. Rather, the fields and the energy that they contain propagate through space at the speed of light. These propagating fields are called **electromagnetic radiation.** If one were to measure the energy passing through a small area of a sphere of radius R centered about the accelerating charge, the energy flux per unit area would be proportional to the square of the fields or $1/R^2$. But it is evident that the energy flux through the entire sphere, whose area is proportional to R^2, will be independent of the distance R from the charge. So the energy associated with the electromagnetic radiation does not diminish with distance. Because of its acceleration, the charge radiates energy that it cannot recover. This radiated energy forms the basis of all electromagnetic communication over large distances: radio and television, visible light, radar, microwave transmission, etc.

Example 12. Electromagnetic Radiation from an Oscillating Charge

An accelerating charge that is of special interest is one that is oscillating with simple harmonic motion of frequency f. Suppose the charge is oscillating along the z-axis and we wish to compute **E** and **B** at a point on the K^{x-}axis. From Equation 26 we have

$$\mathbf{E} = \frac{q}{4\pi\epsilon_0 c^2} \frac{a'}{x} \mathbf{k}$$

where a′ is the acceleration at time (t − R/c) and **k** is the unit vector in the z-direction. If the displacement of the charge is

$$z = A \sin 2\pi ft$$

then the acceleration is

$$a = \frac{d^2z}{dt^2} = -A\, 4\pi^2 f^2 \sin 2\pi ft$$

Lumping constant factors into the amplitude E_o, we have for the electric field

$$\mathbf{E} = E_0 \sin 2\pi f(t - R/c)\, \mathbf{k} \qquad (27)$$

Similarly,

$$\mathbf{B} = B_0 \sin 2\pi f(t - R/c)\, \mathbf{j}$$

where **j** is the unit vector in the y-direction.

Thus, the fields of the electromagnetic radiation are sinusoidal functions of both distance and time that oscillate at the same frequency as the charge. Just as the period is the time needed to go through one complete cycle, we define the wavelength λ as the distance needed for one complete cycle:

$$\lambda = c/f$$

Classical electromagnetic theory postulates that all electromagnetic radiation is ultimately produced by accelerating charges, travels at the speed of light, and differs only in its frequency of oscillation.

IV. MODERN PHYSICS

The great theories of classical physics such as Newtonian dynamics, electromagnetism, and thermodynamics performed superbly in describing most physical phenomena observed before the 20th century. However, experimental discoveries made at the turn of the century forced extraordinary revisions to be made, giving rise to what is called "modern physics": the theory of relativity and the quantum theory.

Einstein's theory of special relativity is a reconsideration of how terms such as length, time, and mass are defined and is essential to the description of objects moving at very great speeds. It retains many of the underlying assumptions of classical mechanics, such as the idea that all observables can be measured with arbitrary accuracy at the same time. The quantum theory of physics on the other hand is radically different from classical physics and has been extremely successful in describing the behavior of systems of atomic dimensions. It abandons the idea of calculating precisely the trajectories of particles and deals instead with the prediction of the probability of occurrence of events. Both the quantum theory and the theory of special relativity supplant classical physics in the sense that they are capable of accurately predicting physical phenomena in all situations, so far as is known.

A. Special Relativity

1. Galilean and Lorentz Transformations

We recall from Section III.A.2 that Newton's laws of motion are valid only when the coordinate system in which the position, velocity, and acceleration of objects are measured is an inertial system, i.e., one in which an object that is not acted on by any force appears to move with constant velocity. It follows that any coordinate system moving at constant velocity with respect to an inertial system is itself an inertial system. One might imagine for example a laboratory inside a space ship that is drifting at constant speed far from any external influences. Newton postulated that to an experimenter in an inertial system the laws of physics are always the same regardless of which inertial system he is in. In particular, if experiments were carried out in two inertial systems that are moving relative to each other with constant velocity v, it would not be possible by any experiment to distinguish the one system from the other. Suppose the coordinate axes for each system are parallel and that the relative motion is along the x-direction. If we wish to write the expressions relating the coordinates in the one system to those in the other, we would probably assume that the correct **transformation equations** are

$$x_2 = x_1 - v\,t_1$$

$$y_2 = y_1$$

$$z_2 = z_1$$

$$t_2 = t_1$$

if at zero time their origins coincided. These are called the **Galilean coordinate transformations.** It is a straightforward task to show that principles such as the conservation of linear momentum and of energy retain their mathematical form when coordinates are altered in this way.

Problems arose when these simple ideas were applied to the laws of electromagnetism. These laws, known as Maxwell's equations, do change their form under a Galilean transformation of coordinates, implying that electromagnetic phenomena will appear differently to observers moving at different velocities. In particular, it was predicted that the measured velocity of light, like the measured velocity of any object, should depend on the velocity of the coordinate system in which it is measured. However, the startling results of the famous Michelson-Morley experiment suggested that the speed of light appears the same regardless of the velocity of the observer.

Einstein proposed that difficulties could be resolved (1) by retaining the postulate that all laws of physics be independent of the motion of the inertial coordinate system to which they are referred and (2) by adding a second postulate that the velocity of light have the same value in all inertial systems. If the laws of electromagnetism are to remain invariant in all inertial systems then apparently the Galilean transformation is incorrect.

A correct set of transformation equations can be derived by using Einstein's postulates along with two simple assumptions. The first assumption is that the equations of transformation are linear; otherwise a coordinate in one system might correspond to two or more coordinates in a second system. Second, we assume as with the Galilean transformations that coordinates in directions perpendicular to the direction of the relative velocity are the same in both systems. Suppose we have two coordinate systems moving with relative velocity v with respect to each other along the direction of their x-axes. If at time $t = 0$, when their origins coincide, a burst of light is emitted from the origin of their coordinate systems, a sphere of light will radiate outwards with a radius equal to the speed of light c multiplied by the time. For System 1, any point (x_1, y_1, z_1) on the sphere is given by

$$x_1^2 + y_1^2 + z_1^2 = c^2 t_1^2$$

and similarly for System 2. If the speed of light is to be the same in both coordinate systems, then

$$x_1^2 + y_1^2 + z_1^2 - c^2 t_1 = x_2^2 + y_2^2 + z_2^2 - c^2 t_2^2$$

or

$$x_1^2 - c^2 t_1^2 = x_2^2 - c^2 t_2^2 \qquad (28)$$

where we have used the assumption that $y_1 = y_2$ and $z_1 = z_2$. Now, the origin of System 2 has an x-coordinate in System 1 of $x_1 = vt_1$ or $x_1 - vt_1 = 0$. But in System 2 the coordinate of the origin is always $x_2 = 0$. Since these two equations describe the same fact,

$$x_2 = k(x_1 - vt_1) \qquad (29)$$

where k is a constant since we have assumed a linear relationship between coordinates. Similarly, for the origin of System 1 we have the same relation with v replaced by −v:

$$x_1 = k(x_2 + vt_2) \qquad (30)$$

The value of k must be the same in Equations 29 and 30 if the laws of physics are not going to depend on the velocity of the inertial system. Using Equations 28 to 30, one arrives at the following relations:

$$x_2 = \frac{x_1 - vt_1}{\sqrt{1 - v^2/c^2}}$$

$$y_2 = y_1$$

$$z_2 = z_1$$

$$t_2 = \frac{t_1 - vx_1/c^2}{\sqrt{1 - v^2/c^2}}$$

These are called the **Lorentz transformation equations** and for small velocities v they reduce to the Galilean relations. Happily, it turns out that all physical laws, including those of electromagnetism, remain the same under a Lorentz coordinate transformation.

Several unexpected consequences contrary to everyday experience follow from the Lorentz transformation equations. These consequences are a result of the fact that when an observer makes time and length measurements on a system that is moving with respect to his own, the measurements must involve the transmission of signals between several locations at a speed which cannot exceed the speed of light. Only when an observer or the system that he is observing is moving at a speed approaching that of light, however, do these unfamiliar effects become noticeable.

2. Simultaneity

If two events occur at different locations in the coordinate system of an observer, he can define them to have happened simultaneously if light signals emitted when the events occurred arrive at the geometrically measured midpoint between the two locations at the same time. However, two events that are simultaneous in his system will not be simultaneous when viewed from another system. This is because an observer moving with velocity v with respect to the first system will see the one signal before the other, since he is moving during the finite time required for the light to reach him.

Suppose two events occur at x_1 and x_1' in System 1 at times t_1 and t_1', respectively. From the Lorentz transformation equations we see that if $t_1 = t_1'$ for System 1, then for System 2

$$t_2 = \frac{t_1 - \frac{vx_1}{c^2}}{\sqrt{1 - \frac{v^2}{c^2}}}$$

and

$$t_2' = \frac{t_1 - \frac{vx_1'}{c^2}}{\sqrt{1 - \frac{v^2}{c^2}}}$$

Hence, t_2 and t_2' are not the same unless the events occurred at the same location $x_1 = x_1'$.

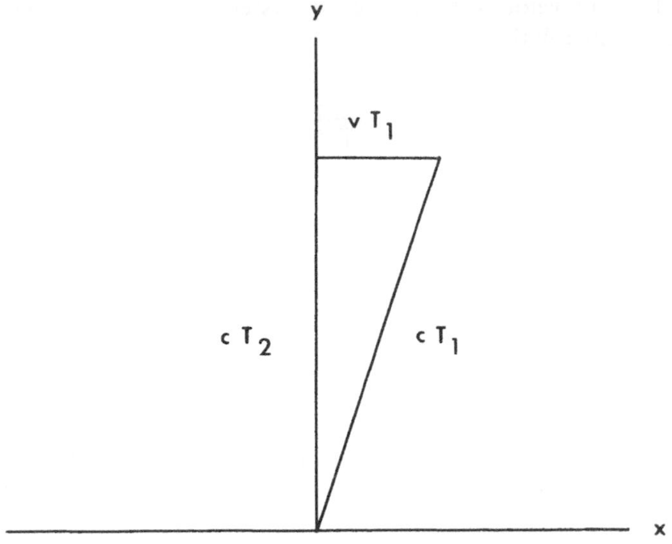

FIGURE 10. Relationship between the distance cT_2 traveled by a burst of light shot along the y-axis of System 2 and the distance cT_1 traveled by the light as measured in System 1. System 2 is moving to the right at velocity v relative to System 1.

3. Time Dilation

Suppose we look at System 1 as being at rest and System 2 moving with respect to it at velocity v along their common x-axes. A person in System 2 shooting a burst of light along the y-axis of System 2 will find that in time interval T_2 the light will travel a distance $y_2 = cT_2$. We choose the y-axis because position coordinates perpendicular to the direction of motion appear the same for all coordinate systems. Now, to an observer in System 1 the light pulse appears to travel obliquely upwards because System 2 moves a distance vT_1 in the x direction. The apparent distance the light moves in System 1 is cT_1 and by the Pythagorean theorem (see Figure 10)

$$c^2T_1^2 = c^2T_2^2 + v^2T_1^2$$

or

$$T_1 = \frac{T_2}{\sqrt{1 - \frac{v^2}{c^2}}}$$

Thus, to the observer in System 1 the time interval appears longer than it does to the observer in System 2; as though events in the moving system were proceeding more slowly.

4. Length Contraction

If a rod positioned along the x-axis of System 2 has length L_2 as measured in that system, we wish to know what length it will appear to have in System 1. Suppose the observer in System 1 notices that it takes a time T_A for the rod to pass by him. Then the apparent length of the rod is

$$L_1 = vT_A$$

Now, the equivalent length of time in System 2 is the interval

$$T_B = \frac{T_A}{\sqrt{1 - \frac{v^2}{c^2}}}$$

and represents the time it appears to take System 1 to pass by the rod. That is, to the observer in System 2

$$L_2 = vT_B = \frac{vT_A}{\sqrt{1 - \frac{v^2}{c^2}}}$$

Eliminating v, we have

$$L_1 = L_2 \sqrt{1 - \frac{v^2}{c^2}}$$

This says that the length of an object moving with respect to an observer appears to be shorter parallel to the direction of motion.

5. Addition of Velocities

Suppose an object has velocity **u** in System 1 and we wish to find its velocity **u′** as seen in System 2. As usual, System 2 is moving with velocity v relative to System 1 along their common x-axes. In System 1 the object has x-, y-, and z-components of velocity

$$u_x = \frac{dx_1}{dt_1}$$

$$u_y = \frac{dy_1}{dt_1}$$

$$u_z = \frac{dz_1}{dt_1}$$

while in System 2 the velocity components are

$$u'_x = \frac{dx^2}{dt_2}$$

$$u'_y = \frac{dy_2}{dt_2}$$

$$u'_z = \frac{dz_2}{dt_2}$$

By differentiating the Lorentz transformation equations, it is straightforward to show that

$$u_x' = \frac{u_x - v}{w}$$

$$u_y' = \frac{u_y \sqrt{1 - \frac{v^2}{c^2}}}{w}$$

$$u_z' = \frac{u_z \sqrt{1 - \frac{v^2}{c^2}}}{w} \qquad (31)$$

where

$$w = 1 - \frac{vu_x}{c^2}$$

The form of these equations is such that the velocity of the object can never appear to exceed the speed of light c.

6. Dependence of Mass on Velocity

Suppose we have an xy-coordinate system in which two observers are moving toward each other along the x-axis with a relative velocity v. As they pass, each projects a particle of the same mass toward the other in a direction which appears to him to be perpendicular to the x-axis and with velocity u. There are three different coordinate systems from which the collision of the particles can be viewed: those of the two observers and the xy-coordinate system. We wish to examine the law of conservation of momentum in each system. From the viewpoint of the xy-coordinate frame, the particles appear to approach the x-axis at the same angle θ and with the same speed so that their total vector momentum before the collision is zero. After the collision, if the law of conservation of momentum is to be valid, the particles must leave in exactly opposite directions and with the same speeds. Any angle of departure from the x-axis is possible but let us suppose for simplicity that this is the same angle θ at which the particles approached. From the viewpoint of one of the observers, his particle arrived along a direction parallel to the y-axis and returned along the same line with the same speed u. To him, the other particle arrived at a shallow angle to the x-axis and departed at the same angle without changing speed.

Now, to the observer who projected particle 1, although the total x-momentum of the particles is not zero because of the apparent x-component of velocity of particle 2, the total y-momentum of the particles is zero. This is because the y-components of the velocities of the two particles appear to him to simply change sign after the collision. To observer 1 the y-component of velocity of particle 2 is given from Equation 31 as

$$u_y' = \frac{u_y \sqrt{1 - \frac{v^2}{c^2}}}{1 - \frac{vu_x}{c^2}}$$

where $u_y = u$ and $u_x = 0$ are the velocity components of particle 2 as seen by observer 2. Then,

$$u'_y = u\sqrt{1 - \frac{v^2}{c^2}}$$

So to observer 1, a zero y-component of momentum implies that

$$m_1 u = m_2 u\sqrt{1 - \frac{v^2}{c^2}}$$

Apparently m_1 does not equal m_2 from the viewpoint of observer 1 and instead

$$m_2 = \frac{m_1}{\sqrt{1 - \frac{v^2}{c^2}}}$$

In the event that the velocity u with which each particle is projected is almost zero, then to observer 1 his particle is almost at rest while particle 2 is moving with approximately the velocity v at which observer 2 is approaching observer 1. Then

$$m_v = \frac{m_0}{\sqrt{1 - \frac{v^2}{c^2}}} \qquad (32)$$

which states that the mass of an object moving with velocity v with respect to an observer appears greater than its mass at rest m_0. Subatomic particles in atoms and accelerating devices are capable of extremely high speeds and their increase in mass has been experimentally verified. The dependence of mass on velocity provides us with an explanation of why an object cannot travel faster than the speed of light. As the velocity of an object approaches that of light, the mass of the object becomes infinitely great so that it experiences almost no acceleration under the action of a force upon it.

7. Equivalence of Mass and Energy

If we expand Equation 32 using the binomial theorem, we have

$$m = \frac{m_0}{\sqrt{1 - \frac{v^2}{c^2}}}$$

$$= m_0\left(1 + \frac{v^2}{2c^2} + \cdots\right)$$

$$\cong m_0 + \tfrac{1}{2} m_0 v^2 \left(\frac{1}{c^2}\right)$$

Multiplying both sides of the equation by c^2, we have

$$mc^2 \cong m_0 c^2 + \tfrac{1}{2} m_0 v^2$$

The terms of this equation have dimensions of energy, $\frac{1}{2} m_0 v^2$ being the classical kinetic energy of a particle. Although an approximation, this equation suggests that the energy of a particle moving with velocity v may be expressed as mc^2, consisting of an intrinsic energy term m_oc^2 due to the rest mass of the object plus the kinetic energy. That is, whereas we had considered the total energy of a particle to be composed of only its potential and kinetic energies, relativity theory makes it apparent that a particle has, in addition, energy due to its mass as measured when the particle is at rest.

The exact expression for the total energy E of a particle is

$$\begin{aligned} E &= mc^2 \\ &= m_0 c^2 + T + U \qquad (33) \end{aligned}$$

where T is the kinetic energy of the particle and U is the potential energy. The relativistic kinetic energy T is equal to the increase in the mass of the particle due to its motion multiplied by c^2. That is, if the potential energy of the particle is zero, then

$$T = (m - m_0)\,c^2 = m_0 c^2\,(1/\sqrt{1 - v^2/c^2} - 1)$$

Although T thus defined reduces to $\frac{1}{2}m_ov^2$ for small velocities, it is not true that T is given by $\frac{1}{2}mv^2$ where m is the relativistic mass.

Equation 33 states that any increase or decrease in T or U for an object (or more generally in any kind of energy associated with the object: thermal, chemical, nuclear, etc.) will be accompanied by a corresponding increase or decrease in the mass of the object. One may therefore consider mass and energy to be equivalent and interchangeable so that the law of conservation of energy is broadened to include mass as well as energy.

The magnitude of the relativistic linear momentum **p** is

$$\mathbf{p} = mv = \frac{m_0 v}{\sqrt{1 - \frac{v^2}{c^2}}}$$

With the potential energy taken to be zero, the total energy E is related to p by the formulas

$$E = \frac{pc^2}{v}$$

and

$$E^2 = p^2 c^2 + m_0{}^2 c^4 \qquad (34)$$

B. Quantum Theory

1. Particle Aspects of Electromagnetic Radiation

Although the question of whether electromagnetic radiation is wave-like in character or consists of discrete particles had long been debated by scientists, the question seemed resolved in favor of waves when the interference and diffraction of light were observed and interpreted in the early 19th century. For example, if a beam of light is directed at two narrow parallel slits, the pattern seen on a screen illuminated by the slits is not two lines of light as one would expect if light was completely particle-like. Instead a series of bright and dark bands is seen. These "interference fringes" are predicted by wave theory as resulting from the superimposition of the waves emanating from the slits.

However, the wave-particle dilemma returned in the early 20th century when it became apparent that certain newly discovered phenomena such as the photoelectric effect required for their explanation that electromagnetic radiation be particle-like in nature. In the photoelectric effect, electromagnetic radiation incident on a clean metal surface causes electrons to be emitted from the surface. Experiment showed that the kinetic energy of the emitted electrons depends on the frequency of the radiation rather than on its intensity, i.e., on the energy transported by the radiation per unit time and area. As shown in Section III.B.4, classical wave theory would predict that the electric field of the electromagnetic radiation causes the electrons of the metal surface to oscillate, acquiring an energy proportional to the square of the electric field, i.e., proportional to the intensity of the radiation. Hence, wave theory cannot provide an explanation of the photoelectric effect. However, if as Einstein suggested, electromagnetic radiation consists of energy packets each carrying an energy proportional to the frequency of the radiation, the experimental results can be understood. Then the photoelectric effect is seen as the collision of individual energy packets with individual electrons in which each packet gives up its entire energy to the electron.

From this and other experimental evidence it is now known that a monochromatic beam of electromagnetic radiation, classically described as a wave of frequency ν and wavelength λ, at times behaves as a collection of energy packets called **photons**. Each photon has an energy E that is proportional to the frequency:

$$E = h\nu = \frac{hc}{\lambda} \tag{35}$$

where Planck's constant h is the fundamental constant of the quantum theory and has the value

$$h = 6.626 \times 10^{-34} \quad \text{J/sec}$$

Using Equation 34 supplied by the theory of relativity, we can assign to the photon a value for its linear momentum p:

$$p = \frac{E}{c} = \frac{h\nu}{c} = \frac{h}{\lambda} \tag{36}$$

What we have with electromagnetic radiation is behavior that is neither exactly like that of waves nor exactly like that of particles. A quantity that bridges the gap between the two properties is the intensity I of the radiation, the energy which flows in unit time through unit area perpendicular to the direction of propagation. In classical wave theory the intensity I is proportional to the square of the electric field E (or magnetic field B). In terms of photon flux, I is given by the average number of photons arriving per unit time per unit area multiplied by the energy per photon. Hence, the probability of observing a photon at any point is proportional to E^2. When several beams of radiation superimpose, their fields add and the square of the total field gives the probability of finding a photon at any point.

Example 13. Young's Experiment

In the two-slit (or Young's) experiment mentioned above, a beam of light of frequency ν is incident on two narrow, parallel slits as shown in Figure 11. From each slit emerges an expanding beam of light that falls on a screen a large distance away. The electric fields of the radiation from each slit are given by Equation 27:

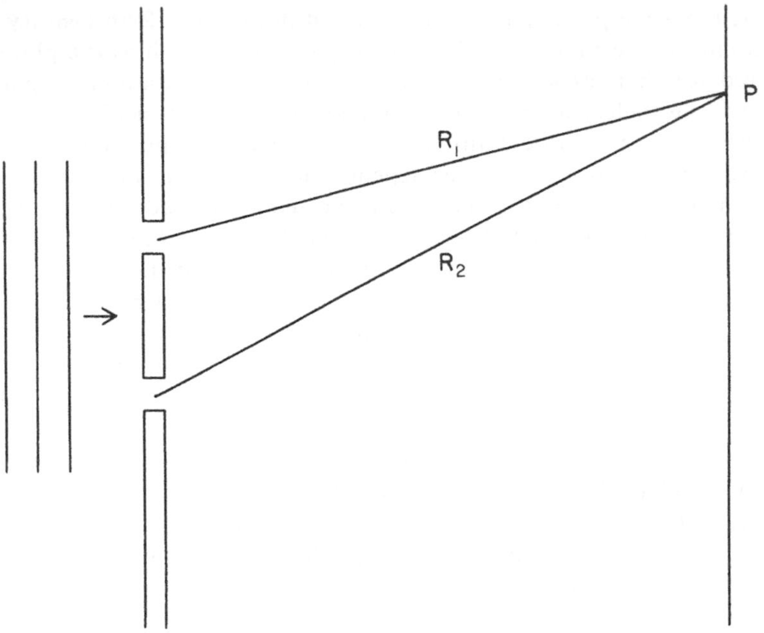

FIGURE 11. Young's experiment, in which beams of light emanating from two parallel slits combine at a point on a screen.

$$E_1 = E_o \sin \omega \left(t - \frac{R_1}{c}\right)$$

$$E_2 = E_o \sin \omega \left(t - \frac{R_2}{c}\right)$$

where R_1 and R_2 are the distances from the slits to a point on the screen and $\omega = 2\pi\nu$. Where the fields superimpose, the resultant field is the sum of E_1 and E_2:

$$\begin{aligned} E_{total} &= E_1 + E_2 \\ &= E_o \left[\sin \omega \left(t - \frac{R_1}{c}\right) + \sin \omega \quad t - \frac{R_2}{c}\right)\right] \\ &= 2\, E_o \cos \omega \left(\frac{R_2 - R_1}{2c}\right) \sin \omega \left(t - \frac{R_1 + R_2}{2c}\right) \end{aligned}$$

where we have used the relation

$$\sin A + \sin B = 2 \cos \tfrac{1}{2} (A - B) \sin \tfrac{1}{2} (A + B)$$

The intensity I_p of the radiation at point P, which is proportional to E^2_{total}, is then

$$I_p = I_o \sin^2 \omega \left(t - \frac{R_1 + R_2}{c}\right) \qquad (37)$$

where

$$I_o = 4 E_o^2 \cos^2 \omega \left(\frac{R_2 - R_1}{2c} \right)$$

Now, I_o gives the maximum intensity that can be observed at point P since the sine term of Equation 37 simply modulates the intensity as a function of time. We see that the intensity at any point of the screen varies as the square of the cosine of the difference between R_2 and R_1.

In terms of photons, the number N arriving per unit time and area of the screen is proportional to the intensity divided by the energy per photon:

$$N \propto \frac{I_p}{h\nu}$$

Remarkably, interference fringes occur even when the intensity of the light beam is reduced to such an extent that individual photons pass through one at a time. Interference effects, rather than being the result of photons interfering with each other, are related to the fact that the probability of finding a photon at a particular location is given by the square of a wave-like function.

2. Wave Aspects of Particles

It is remarkable to those of us who have had experience only with large-scale objects that the behavior of very small particles is in some respects wave-like. In fact, there is an exact analogy between the wave-particle duality of electromagnetic radiation and the wave-particle duality associated with electrons, protons, neutrons, and the other constituents of matter. If the two-slit experiment described in Example 13 for electromagnetic radiation were to be repeated using electrons instead, the same kind of interference pattern would be observed. Actually, to show these effects with electrons requires slits too small to mechanically construct, but the scattering of an electron beam from the ordered atoms of a crystal is feasible and the experiment was performed by Davisson and Germer in 1927. The experiment showed that particles of energy E and momentum p behave like waves of frequency

$$\nu = E/h \tag{38}$$

having a wavelength, known as the **de Broglie wavelength** of

$$\lambda = h/p \tag{39}$$

These relations are identical to those for the photon (Equations 35 and 36).

Just as an electromagnetic wave is specified by an electric and magnetic field whose square gives the probability of finding a photon at a particular point, so for material particles there is an associated **wave function** Ψ whose square is proportional to finding the particle at a point in space. There is an apparent difference between electromagnetic and matter waves, viz., the wave functions happen to have real and imaginary parts whereas the fields E and B of electromagnetism are real quantities. However, the important point is that two variables are required to completely describe the waves in both cases; E and B for electromagnetic waves and the real and imaginary parts of Ψ for the particle wave function.

Calling f and g the real and imaginary parts of Ψ, we have

$$\Psi = f + ig$$

where $i = \sqrt{-1}$. The complex conjugate of a complex number such as Ψ, denoted by Ψ^*, is found by changing the sign of i wherever it appears:

$$\Psi^* = f - ig$$

To make the probability interpretation, we find the absolute square of Ψ, denoted $|\Psi|^2$, which is given by the product of Ψ with its complex conjugate.

$$\begin{aligned}\text{Probability of finding particle in volume dV} &= \Psi\,\Psi^*\,dV \\ &= |\Psi|^2\,dV \\ &= (f + ig)\,(f - ig)\,dV \\ &= (f^2 + g^2)\,dV\end{aligned}$$

Ψ must be normalized so that the probability of finding the particle somewhere in space is unity. This condition is specified by integrating $|\psi|^2$ over all space:

$$\int |\Psi|^2 dV = 1 \quad (40)$$

As an example, we consider the two-slit experiment of Example 13 with the beam of light replaced by an electron beam. Since there are two ways in which an electron may reach the screen or detector, there is a wave function for each. Suppose Ψ_1 is the wave function associated with detecting an electron that has passed through slit 1 while Ψ_2 is the wave function for an electron passing through slit 2. Then the probability P of detecting an electron at any point is given by the absolute square of the sum of the two wave functions:

$$P = |\Psi_1 + \Psi_2|^2$$

As will be seen in Section IV.B.4, Ψ_1 and Ψ_2 are sinusoidal functions of time and position so that interference effects exactly analogous to those for light are found.

3. Heisenberg Uncertainty Principle

Since the probability of finding a particle at a point in space is related to a wave-like function Ψ and since a wave by its very nature is spread to some extent throughout space, it is reasonable to expect that there will be some uncertainty associated with the precise location of the particle. For example if Ψ is a sine wave of wavelength λ, then it extends indefinitely throughout space. Ψ will be non-zero essentially everywhere, implying that the position of the particle is not localized and is unknown. On the other hand in this case we know the wavelength λ of the particle exactly and so, from Equation 39

$$p = h/\lambda$$

we do know the momentum p of the particle exactly. At the other extreme, it is possible by superimposing many sinusoidal waves of various wavelengths to construct a wave function that is highly localized in space. Such a wave packet corresponds to a particle whose position is known with some certainty. However, since waves of many wavelengths were necessary to construct the packet, there will be a large uncertainty in the momentum of the particle.

Hence, we arrive at the law that the position and momentum of a particle cannot be simultaneously known exactly. Heisenberg stated this relationship more precisely. For a particle moving in the x-direction, if the uncertainty in its x-location is Δx and the uncertainty in its x-momentum is Δp_x, then the product of the two is greater than or equal to Planck's constant divided by 2π:

$$\Delta x \, \Delta p_x \geqq \frac{h}{2\pi}$$

It is evident that decreasing the uncertainty of one quantity can only increase the uncertainty of the other. Similar relations exist for other pairs of physical quantities: the uncertainty in the energy ΔE of a particle and the time interval Δt during which the energy measurement was made; and the uncertainty in the simultaneous measurement of any two components of angular momentum, e.g., ΔL_x and ΔL_y.

From the standpoint of the measurement itself, Heisenberg's uncertainty principle can be understood in the following way. In dealing with objects of extremely small dimensions, the effect of measuring one quantity is to disturb the other so as to make its value uncertain. No degree of experimental care or cleverness can avoid this consequence. For example, if one wished to locate an electron precisely he might attempt to scatter a photon or other particle from it. But in the collision the electron would recoil, completely altering its original state of motion so that its momentum would be left uncertain.

4. Schrödinger's Equation

Let us now examine how the wave function Ψ of a particle can be determined for any specific situation. The equation for Ψ was discovered by Schrödinger in 1926. The Schrödinger equation, which is similar to certain wave equations of classical physics, cannot be derived although arguments can be made for its plausibility. We state it here without justification.

In the most general case, Ψ will have different values for different points in space and for different times. That is, Ψ is a function of x, y, z, and t. Suppose a particle has mass m and potential energy U. In general, U is also a function of x, y, z, and t. Then Schrödinger's equation is

$$-\frac{\hbar^2}{2m}\left(\frac{\partial^2\Psi}{\partial x^2} + \frac{\partial^2\Psi}{\partial y^2} + \frac{\partial^2\Psi}{\partial z^2}\right) + U\Psi = i\hbar\frac{\partial\Psi}{\partial t}$$

where

$$\hbar = h/(2\pi)$$

This is a differential equation involving first and second partial derivatives. The solution of the equation gives the wave function Ψ of the particle.

Under certain circumstances, Schrödinger's equation may be substantially simplified. First, let us suppose that the particle is confined to move in one dimension, say x. Then the partial derivatives with respect to y and z are zero and Schrödinger's equation becomes

$$-\frac{\hbar^2}{2m}\frac{\partial^2\Psi}{\partial x^2} + U\Psi = i\hbar\frac{\partial\Psi}{\partial t} \qquad (41)$$

A further simplification exists when the potential energy U does not depend explicitly on time. In that case it is possible to use a method called **separation of variables** in which Ψ is separated into a product of two functions, one ψ which depends only on x and the other ϕ which depends only on t:

$$\Psi(x,t) = \psi(x)\,\phi(t)$$

Substituting into Equation 41, we have

$$-\frac{\hbar^2}{2m}\frac{\partial^2(\psi\phi)}{\partial x^2} + U\psi\phi = i\hbar\frac{\partial(\psi\phi)}{\partial t}$$

$$-\frac{\hbar^2}{2m}\phi\frac{\partial^2\psi}{\partial x^2} + U\psi\phi = i\hbar\psi\frac{\partial\phi}{\partial t}$$

$$-\frac{\hbar^2}{2m}\frac{1}{\psi}\frac{\partial^2\psi}{\partial x^2} + U = i\hbar\frac{1}{\phi}\frac{\partial\phi}{\partial t}$$

The left side of the equation depends only on the variable x while the right side depends only on t. Since x and t are independent variables, the equation can hold for any arbitrary value of x and t only if both sides are equal to some constant, which we call W. Also, we may replace the partial differentiation by ordinary differentiation since ψ and ϕ are functions of single variables only. We then have two ordinary differential equations:

$$-\frac{\hbar^2}{2m}\frac{1}{\psi}\frac{d^2\psi}{dx^2} + U = W \tag{42}$$

and

$$i\hbar\frac{1}{\phi}\frac{d\phi}{dt} = W \tag{43}$$

The solution of Equation 43 is

$$\phi(t) = e^{-\frac{iWt}{\hbar}}$$

which, as seen in Example 4, is an oscillatory function of time of frequency

$$\nu = \frac{W}{2\pi\hbar} = \frac{W}{h}$$

We also know that the frequency of the wave function is related to the energy E of the particle by Equation 38:

$$E = h\nu$$

This indicates that the separation constant W appearing in Equations 42 and 43 is identical to the total energy E of the particle and henceforth the symbol E will be used for it.

As for Equation 42, multiplying both sides by ψ we have

$$-\frac{\hbar^2}{2m}\frac{d^2\psi}{dx^2} + U\psi = E\psi \tag{44}$$

This is called the **time-independent Schrödinger equation** and is the form of Schrödinger's equation used for one-dimensional problems when the potential energy does not depend explicitly on the time.

Let us suppose that U as a function of position is known for a particle. We would expect that for each possible value of the total energy E of the particle a solution ψ of Equation 44 could be found. However, ψ must be a mathematically well-behaved function, finite and continuous, if the probability interpretation of $|\psi|^2$ is to make sense. It happens, as we shall see in some examples to follow, that not all values of E result in physically admissible wave functions. Let us designate those energies that are allowed for the particle as E_n, being symbolic of a number of discrete or continuous values of the total energy. These are called the **eigenvalues** (or characteristic values) of E for the particular potential energy U. The corresponding solutions ψ_n of the time-independent Schrödinger equation are called the **eigenfunctions.** The index n, which designates a particular eigenfunction and eigenvalue, is called a **quantum number.**

Corresponding to each eigenfunction ψ_n there is a wave function $\Psi_n = \psi_n e^{-iE_nt/\hbar}$. It is easy to verify that any linear combination of the Ψ_n is a solution of Schrödinger's equation. For a particle having a time-independent potential energy the wave function Ψ can always be written:

$$\Psi = \sum_n a_n \psi_n e^{-iE_nt/\hbar}$$

where the a_n are constants.

In the simplest case the time-independent part of the wave function of the particle consists of a single eigenfunction ψ_n:

$$\Psi = \psi_n e^{-iE_nt/\hbar}$$

In physical terms this corresponds to the situation in which the energy E_n of the particle is fixed and known. The probability of finding the particle at position x in this case does not vary with time since

$$|\Psi|^2 = (\psi_n e^{-iE_nt/\hbar})(\psi_n e^{iE_nt/\hbar}) = |\psi_n|^2$$

which depends only on position. The particle is said to be in a **stationary state** or **eigenstate** characterized by the quantum number n.

An important property of eigenfunctions can be derived by considering Schrödinger's time-independent equation for two different eigenfunctions ψ_n and ψ_m:

$$-\frac{\hbar^2}{2m}\frac{d^2\psi_n}{dx^2} + U\psi_n = E_n\psi_n \tag{45}$$

$$-\frac{\hbar^2}{2m}\frac{d^2\psi_m^*}{dx^2} + U\psi_m^* = E_m\psi_m^* \tag{46}$$

where the complex conjugate of Equation 45 has been taken. Multiplying Equation 45 by ψ^*_m and Equation 46 by ψ_n and subtracting, we have

$$\frac{2m}{\hbar^2}(E_n - E_m)\psi_m^*\psi_n = \psi_n \frac{d^2\psi_m^*}{dx^2} - \psi_m^* \frac{d^2\psi_n}{dx^2}$$

We now integrate from $x = -\infty$ to $+\infty$:

$$\frac{2m}{\hbar^2}(E_n - E_m)\int_{-\infty}^{\infty} \psi_m^*\psi_n dx = \int_{-\infty}^{\infty} \frac{d}{dx}\left(\psi_n \frac{d\psi_m^*}{dx} - \psi_m^* \frac{d\psi_n}{dx}\right) dx$$

$$= \left[\psi_n \frac{d\psi_m}{dx} - \psi_m^* \frac{d\psi_n}{dx}\right]_{-\infty}^{\infty} \qquad (47)$$

Now, eigenfunctions must go to zero at infinity, otherwise, if one were to calculate the probability of finding the particle anywhere on the x-axis, an infinite number would result. Consequently, the right-hand side of Equation 47 is zero and, if E_n does not equal E_m,

$$\int_{-\infty}^{\infty} \psi_m^* \psi_n dx = 0$$

Functions possessing this quality are said to be **orthogonal**. Combining this result with Equation 40, we have

$$\int_{-\infty}^{\infty} \psi_m^* \psi_n dx = \begin{cases} 1 & \text{if } n = m \\ 0 & \text{if } n \neq m \end{cases} \qquad (48)$$

Under certain circumstances E_m and E_n can be equal even though m and n are different. That is, different eigenfunctions may correspond to the same energy eigenvalue. Such eigenfunctions are said to be **degenerate** and a procedure exists for making such eigenfunctions orthogonal.

Let us now look at several simple examples that will help to illustrate some of the main features of quantum mechanics.

Example 14. Particle in a Box

Consider a particle free from external forces that bounces back and forth between the two walls of a box along the x-axis. Except when it touches a wall it feels no force. We recall (Section III.A.2.h) that the potential energy U of a particle is related to the force F on it by the equation

$$F_x = -\frac{\partial U}{\partial x}$$

Hence, when the particle is not touching a wall and $F = 0$, U must be a constant. Since this is true regardless of the value of the constant, we may as well take $U = 0$ for simplicity. When the particle strikes the wall, it experiences an extremely strong force of short duration. So U must rise sharply from zero at the walls in order that F, which is the slope of U, be large. This situation is illustrated in Figure 12. The walls are located at $x = D/2$ and $x = -D/2$. U is zero between the walls, rising abruptly to

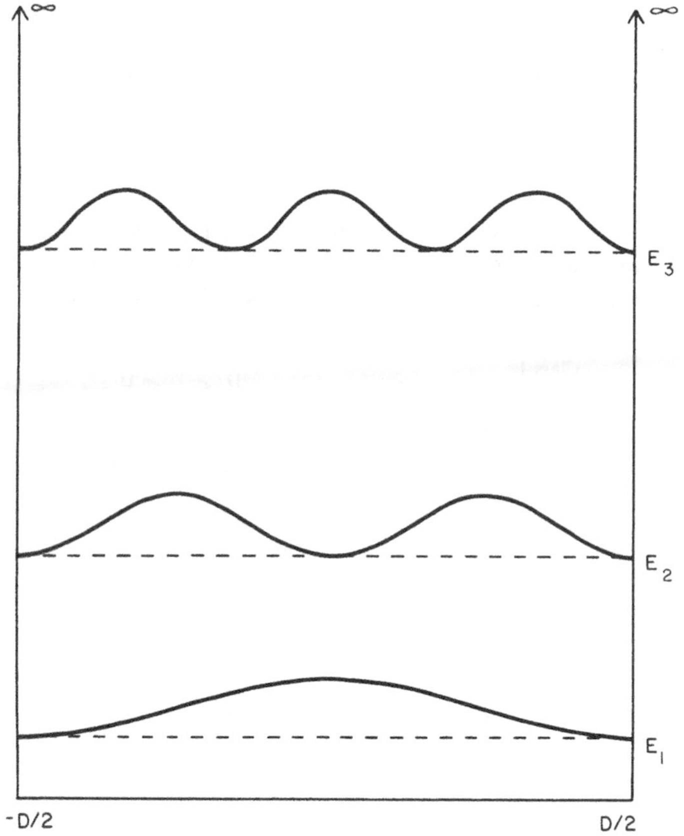

FIGURE 12. The probability densities ψ^2 for the first three eigenstates of a particle in an impenetrable box of width D. The corresponding energy levels (dashed lines) serve as abscissas.

infinity at the walls. The significance of U rising to infinity is that this ensures that the particle will not escape from the box regardless of its energy.

Between the walls, Schrödinger's time-independent equation becomes

$$-\frac{\hbar^2}{2m}\frac{d^2\psi}{dx^2} = E\psi \tag{49}$$

The solution to this equation is

$$\psi = A \sin kx + B \cos kx$$

where

$$k = \sqrt{2mE}/\hbar \tag{50}$$

as can be verified by substituting into Equation 49. According to the probability interpretation of the wave function, the probability of finding the particle at location x is given by $|\psi|^2$. At the walls and outside the box the particle cannot be present. Hence, ψ must vanish at $x = \pm D/2$. Now, the sine function is zero when its argument equals 0, π, 2π, 3π, etc., or when

$$k = 0, \quad 2\pi/D, \quad 4\pi/D, \quad \text{etc.} \tag{51}$$

Similarly, the cosine function vanishes when its argument equals $\pi/2$, $3\pi/2$, etc., or when

$$k = \pi/D, \quad 3\pi/D, \quad 5\pi/D, \quad \text{etc.} \tag{52}$$

However, there is no value of k for which both sin kx and cos kx vanish simultaneously. So we must either set $B = 0$ and choose k so that $\sin kx = 0$ or set $A = 0$ and choose k so that $\cos kx = 0$. That is, there are two classes of solutions:

Class 1

$$\psi_n = \begin{cases} A \sin \dfrac{n\pi x}{D} & \text{for } -\dfrac{D}{2} \leqq x \leqq \dfrac{D}{2} \\ 0 & \text{for } x > \dfrac{D}{2},\ x < -\dfrac{D}{2} \end{cases}$$

where $n = 2,4,6, \cdots$

Class 2

$$\psi_n = \begin{cases} B \cos \dfrac{n\pi x}{D} & \text{for } -\dfrac{D}{2} \leqq x \leqq \dfrac{D}{2} \\ 0 & \text{for } x > \dfrac{D}{2},\ x < -\dfrac{D}{2} \end{cases}$$

where $n = 1,3,5, \cdots$

These, then, are the eigenfunctions for a particle in a box. The numbers n are the quantum numbers that designate each possible eigenfunction. The eigenfunctions in this case are of a kind known as standing waves in which there is no motion of the wave along the x-axis. Standing waves are produced by superimposing two traveling waves of the same wavelength, speed, and amplitude which are traveling in opposite directions. Associating a traveling wave with a moving particle whose direction of motion is known, it is plausible to conclude that a standing wave eigenfunction represents a particle that could be moving in either direction along the x-axis, i.e., a particle whose direction of motion in the box is unknown. However, since the wavelength of each wave function is definite, the magnitude of the momentum of the particle must be known precisely from the de Broglie relation

$$p = \frac{h}{\lambda} = \hbar k$$

The general solution of Schrödinger's equation is

$$\Psi = \psi\,\phi = \psi\, e^{-iEt/\hbar}$$

To find the probability of locating the particle at any particular point inside the box, we must compute $|\Psi|^2$:

$$|\Psi|^2 = (\psi\, e^{-iEt/\hbar})\,(\psi^* e^{iEt/\hbar})$$

$$= |\psi|^2$$

$$= \begin{cases} \dfrac{2}{D}\sin^2\dfrac{n\pi x}{D}, & n = 2, 4, 6, \cdots \\ \dfrac{2}{D}\cos^2\dfrac{n\pi x}{D}, & n = 1, 3, 5, \cdots \end{cases}$$

Several of these probability functions are plotted in Figure 12. Classical physics would say that the particle spends an equal amount of time anywhere along the x-axis inside the box so that its probability of being found is the same everywhere inside. Quantum mechanics, on the other hand, states that there is a wave-like character to this probability and that, surprisingly, points exist where there is zero probability of finding the particle in the box.

Since classical physics is valid only for large-scale phenomena, let us examine the case in which the mass m and the energy E of the particle are large. Then, from Equation 50, k is a large number. As a result, $|\psi|^2$ oscillates so rapidly as a function of position that the oscillations are too closely spaced to be experimentally observed. The average value of $|\psi|^2$ turns out to be just the probability that is predicted by classical theory. That the quantum theory must in all cases yield the same results when applied to large-scale phenomena as does Newtonian mechanics is known as the **correspondence principle.**

Another unusual result is that only certain energies are possible for the particle. From Equations 50 to 52 we have

$$k = \frac{n\pi}{D} = \frac{\sqrt{2mE}}{\hbar}$$

Hence,

$$E_n = \frac{\hbar^2 n^2 \pi^2}{2mD^2}, \qquad n = 1, 2, 3, \cdots$$

In the terminology of quantum mechanics we say that the energy is **quantized,** i.e., that it can only have certain values. These allowed values are termed the energy eigenvalues for the particle. Again, classical physics would not have predicted this result since any kinetic energy would seem acceptable. For particles and boxes of macroscopic size, the spacing between successive values of E_n is infinitesimal so that the energy quantization cannot be observed experimentally.

Example 15. The Square Well Potential

Suppose the potential energy of the particle does not rise infinitely high at the walls but only to some finite height U_0, as shown in Figure 13. This is called a square well potential and is useful in modeling certain atomic and nuclear phenomena. The total energy E of a particle is the sum of its kinetic energy T and its potential energy U. Hence, we have for the kinetic energy

$$T = E - U$$

Now, a particle with total energy E that is less than U_0 cannot exist outside the "well"

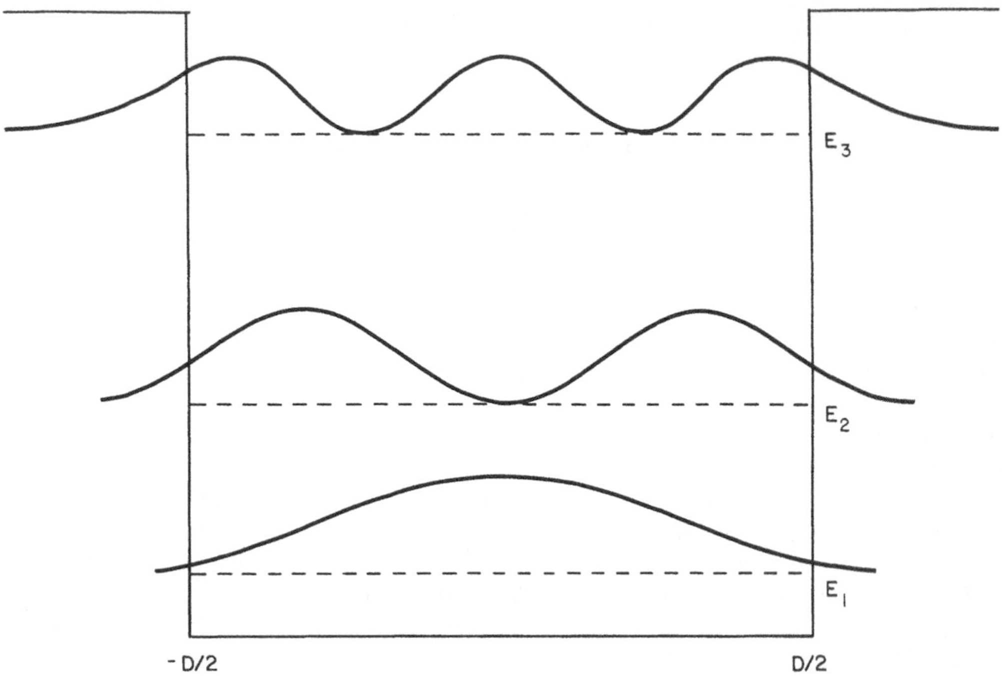

FIGURE 13. The probability densities ψ^2 for the first three eigenstates of a particle in a square well of width D. The corresponding energy levels (dashed lines) serve as abscissas. The "penetration" of the probability densities of the particle beyond the sides of the well is evident.

according to classical physics. This is because the kinetic energy, $T = \frac{1}{2}mv^2$, must be a positive number. When E is greater than U_o however, the particle may occupy any region in or out of the well.

Let us see what results quantum mechanics gives for the case in which the total energy E of the particle is less than the potential well height U_o. Inside the well ($-D/2 < x < D/2$) where the potential energy is zero, the time-independent Schrödinger equation is identical to that for the particle in a box, Equation 49. Outside the well the time-independent Schrödinger equation is

$$-\frac{\hbar^2}{2m}\frac{d^2\psi}{dx^2} + U_o\psi = E\psi$$

The solutions again fall into two distinct classes:

Class 1:

$$\psi = \begin{cases} -A_1 e^{k_1 x}, & x < -\frac{D}{2} \\ A_2 \sin k_2 x, & -\frac{D}{2} < x < \frac{D}{2} \\ A_1 e^{-k_1 x}, & x > \frac{D}{2} \end{cases}$$

Class 2:

$$\psi = \begin{cases} B_1\, e^{k_1 x}, & x < -\dfrac{D}{2} \\ B_2 \cos k_2 x, & -\dfrac{D}{2} < x < \dfrac{D}{2} \\ B_1\, e^{-k_1 x}, & x > \dfrac{D}{2} \end{cases}$$

where

$$k_1 = \frac{\sqrt{2m(U_0 - E)}}{\hbar} \tag{53}$$

and

$$k_2 = \frac{\sqrt{2mE}}{\hbar}$$

The As and Bs are constants which can be found from the normalization condition, Equation 40.

We notice that outside the well, where classically the particle would not be allowed, there is a non-zero wave function that is proportional to (for positive x)

$$e^{-\sqrt{2m(U_0 - E)}x/\hbar}$$

Hence, there exists a finite probability of finding the particle outside the well and this probability is an exponentially decreasing function of distance from the well. The fact that particles of subatomic dimensions can "tunnel through" a classically insurmountable barrier provides an explanation for such phenomena as alpha decay in nuclear physics and tunnel diodes in electronics. There is no violation of the principle of energy conservation here. The tunneling is simply a manifestation of Heisenberg's uncertainty principle that an attempt to localize the particle will lead to an uncertainty in its momentum which may increase its energy above U_0.

As is the case for the particle in a box, only certain possible energies are allowed for the particle in the square well when the energy of the particle is less than the well height. In fact it can be shown by a straightforward analysis of Schrödinger's equation that whenever a particle is in a region in which according to classical physics it would be confined, then its total energy can have only special discrete values. For the square well, these energy eigenvalues cannot be simply expressed and must be obtained by numerical methods.

For an energy E greater than U_0, the solutions ψ of Schrödinger's time-independent equation are sinusoidal functions both inside and outside the well since k_1 of Equation 53 is then a purely imaginary number. Inside the well the wavelength of ψ is $h/\sqrt{2mE}$ while outside it is $h/\sqrt{2m(E-U_0)}$. The shorter wavelength inside corresponds to a higher momentum as would be expected classically. Any energy E whatsoever is allowed for the particle in this case. Again, this is a general result: whenever a particle has a total E greater than its potential energy U in a region of infinite extent, then any total energy is allowed. We say there exists a **continuum** of values for the total energy. The energy levels of a square well potential are illustrated in Figure 13.

5. Operators and Expectation Values

As we have seen, the solutions Ψ of the Schrödinger equation yield the probability of finding a particle at a particular location. Furthermore, in solving Schrödinger's equations we determine the permissible energy values that the particle may have. Actually, considerably more information about the particle can be obtained from Ψ and in fact Ψ determines everything that can be known about the system. The method of obtaining this information will now be explored.

A mathematical operation such as multiplication or differentiation can be represented by a symbol called an **operator**. For example, the operator for differentiation with respect to the variable x is written d/dx. In quantum mechanics it is possible to associate an operator with every physical quantity (position, linear momentum, energy, etc.). That this is so and that the operators have the forms that they do will simply be postulated here, although it is possible to argue for their plausibility. The association between the physical quantity and the operator is simply found. First, one writes the quantity in terms of the coordinates x, y, and z and the components of linear momentum p_x, p_y, and p_z. Then, leaving the coordinate variables unchanged, one replaces p_x, p_y, and p_z with $-ih(\partial/\partial x)$, $-ih(\partial/\partial y)$, and $-ih(\partial/\partial z)$, respectively. For example,

Position:

$$x \rightarrow x$$

Linear momentum:

$$p_x \rightarrow -i\hbar \frac{\partial}{\partial x}$$

Angular momentum:

$$L_x = yp_z - zp_y \rightarrow -i\hbar \left(y \frac{\partial}{\partial z} - z \frac{\partial}{\partial y} \right)$$

$$E = \frac{p^2}{2m} + V(x,y,z)$$

Total energy:

$$\rightarrow -\frac{\hbar^2}{2m} \left(\frac{\partial^2}{\partial x^2} + \frac{\partial^2}{\partial y^2} + \frac{\partial^2}{\partial z^2} \right) + V(x,y,z)$$

The total energy written in operator form is often called the **Hamiltonian**.

Now, suppose we wish to find the average value of, say, the x-position of a particle having an associated wave function Ψ. We know that the probability of finding the particle at time t between x and x + dx is Ψ* Ψ dx. If we were to construct many identical systems and measure the probability of finding the particle at all possible values of x, we could calculate the mean x-value by summing the products of each probability with the value of x:

$$\bar{x} = \int_{-\infty}^{\infty} x \, \Psi^* \, \Psi dx$$

An identical procedure could be used to find the mean value of any function of x relating to the particle. A fundamental postulate of quantum mechanics is an extension of this reasoning: the expected mean of a sequence of measurements of the physical quantity Q in a system with associated wave function Ψ is given by

$$\overline{Q} = \int \Psi^* Q_{op} \Psi \, dx$$

where Q_{op} is the operator corresponding to Q. $\overline{Q}$ is often called the **expectation value** of Q.

As an example, let us compute the expectation value of the energy of a particle having a potential energy that depends only on x. For simplicity let us also assume that the wave function Ψ of the particle has as its time-independent part a single eigenfunction ψ_n of the energy. Then

$$\Psi = \psi_n e^{-iE_n t/\hbar} \tag{54}$$

and

$$\overline{E} = \int_{-\infty}^{\infty} \Psi^* E_{op} \Psi \, dx$$

$$= \int_{-\infty}^{\infty} \psi^* e^{iE_n t/\hbar} \left(-\frac{\hbar^2}{2m} \frac{d^2}{dx^2} + U \right) \psi_n e^{-iE_n t/\hbar} \, dx$$

$$= \int_{-\infty}^{\infty} \psi_n^* \left(-\frac{\hbar^2}{2m} \frac{d^2\psi_n}{dx^2} + U\psi_n \right) dx$$

But the expression in parentheses is the left-hand side of the time-independent Schrödinger equation and equals $E_n \psi_n$. Hence,

$$\overline{E} = \int \psi_n^* E_n \psi_n \, dx$$

$$= E_n \int \psi_n^* \psi_n \, dx$$

$$= E_n$$

This is hardly a surprising result, but it shows that the choice of $-i\hbar\partial/\partial x$ for p_x is a reasonable one. Not only is the expectation value of the energy equal to E_n when the wave function has the simple form of Equation 54, but it can also be shown that the theoretical standard deviation in the measurement of $\overline{E}$ is zero. That is, the measured energy of the system can only have the precise value E_n in that case.

An extension of this line of reasoning leads to the idea of each operator having its own eigenvalues and these eigenvalues being equivalent to the measured values of the corresponding physical quantity. The postulate is stated: the only possible values q_n which a measurement of a physical quantity Q can produce must be found from the equation

$$Q_{op}\Phi_n = q_n\Phi_n \tag{55}$$

where Q_{op} is the operator corresponding to Q. An equation of this type is called an eigenvalue equation. The functions Φ_n are called the eigenfunctions of the operator Q_{op} and the constants q_n are called the eigenvalues.

We are already familiar with the eigenvalue equation for the energy operator; it is the time-independent Schrödinger equation. Another equation of interest is that of the linear momentum which has the eigenvalue equation

$$-i\hbar \frac{d\Phi_n}{dx} = p_n\Phi_n$$

Solutions are of the form

$$\Phi_n = A\, e^{ip_n x/\hbar}$$

In order for the momentum eigenfunctions to be finite everywhere, a condition that all eigenfunctions must obey, p_n must be a real number. Except for this restriction, however, p_n can have any value whatever.

The wave function Ψ associated with a particle may or may not be an eigenfunction of other operators. Suppose Ψ is indeed the eigenfunction Φ_n of Q_{op}. Then a measurement of Q is certain to give the results q_n. If Ψ is not an eigenfunction of Q_{op}, a measurement of Q will yield one q_n as a result, but it will not be possible to predict which one. The very act of making the measurement will change the wave function of the system from Ψ to Φ_n.

As an example, suppose the wave function Ψ of a particle is not composed of a single eigenfunction but of a linear combination of many eigenfunctions:

$$\Psi = \Sigma a_n \psi_n e^{-iE_n t/\hbar}$$

where the a_n are constants. Then Ψ is not an eigenfunction of the energy operator E_{op} because the equation

$$\begin{aligned} E_{op}\Psi &= \Sigma\, a_n E_{op} \psi_n e^{-iE_n t/\hbar} \\ &= \Sigma\, a_n E_n e^{-iE_n t/\hbar} \end{aligned}$$

does not have the form of Equation 55. When we compute the expectation value of the energy in this case, we find

$$\begin{aligned} \bar{E} &= \int \Psi^* E_{op} \Psi\, dx \\ &= \int \left(\sum_n a_n^* \psi_n^* e^{iE_n t/\hbar} \right) E_{op} \left(\sum_m a_m \psi_m e^{-iE_m t/\hbar} \right) dx \\ &= \int \sum_n \sum_m a_n^* a_m E_m \psi_n^* \psi_m e^{iE_n t/\hbar} e^{-iE_m t/\hbar} dx \\ &= \sum_n \sum_m a_n^* a_m E_m e^{i(E_n - E_m)t/\hbar} \int \psi_n^* \psi_m\, dx \end{aligned}$$

But from Equation 48 we know that unless n and m are equal, the integral is zero. Therefore,

$$\bar{E} = \sum_n a_n^* a_n E_n$$

This result states that the expectation value of the energy will not in general be equal to one of the energy eigenvalues but will depend on the values of the coefficients a_n. Although any single measurement of the energy will yield a result equal to one of the energy eigenvalues E_n, a large number of different measurements will give an average over all energy values. From this standpoint, the quantity $a_n^* a_n$ is seen to be the probability that the particle will be found in eigenstate n.

6. Transition Probabilities

We have not yet considered how a system having a definite energy can make a transition to a different energy state. This process is of great significance since it is when atomic or nuclear systems change their energy that radiation is produced. In quantum mechanical terms, the wave function of a system may change when the potential energy varies with time. Solving Schrödinger's equation in this case is usually accomplished by an approximation method known as a **time-dependent perturbation** calculation. The time-dependent portion of the potential energy is considered to be a minor variation, or perturbation, from the time-independent part. This perturbation produces a flow of probability from the system being in the original state to being in another state. The result can be a change in the expectation value of the energy of the system, implying that energy has left or entered it. Or the expectation value of the angular momentum or of other physical quantities may change. The perturbation in general consists of some interaction of the system with its environment, such as the electromagnetic interaction between an electric field and an atom.

Suppose a system that has experienced no time-dependent perturbation is in a definite energy state E_m corresponding to the wave function Ψ_m. As usual, we can write Ψ_m in terms of its time-dependent and time-independent parts:

$$\Psi_m = \psi_m e^{-iE_m t/\hbar}$$

Suddenly at $t = 0$ the perturbation v in the potential energy appears so that the initial potential energy V_o becomes

$$V' = V_0 + v$$

We anticipate that the wave function of the system may change to become Ψ'. Now, it can be shown that the new wave function (or practically any function) can be written as a sum of all the possible wave functions Ψ_n for the system before $t = 0$:

$$\Psi' = \sum_n a_n \Psi_n \qquad (56)$$

Here, the a_n are coefficients that change with time and, if they can be calculated, the new wave function can be known at any time from Equation 56.

Substituting Equation 56 into Schrödinger's equation, we have

$$-\frac{\hbar^2}{2m}\frac{\partial^2\Psi'}{\partial x^2} + V'\Psi' = i\hbar\frac{\partial\Psi'}{\partial t}$$

$$-\frac{\hbar^2}{2m}\frac{\partial^2}{\partial x^2}(\Sigma\, a_n\Psi_n) + (V_0 + v)\,\Sigma\, a_n\Psi_n = i\hbar\frac{\partial}{\partial t}(\Sigma\, a_n\Psi_n)$$

$$\Sigma\, a_n\left(-\frac{\hbar^2}{2m}\frac{\partial^2\Psi_n}{\partial x^2} + V_0\Psi_n - i\hbar\frac{\partial\Psi_n}{\partial t}\right) + \Sigma\, a_n v\Psi_n - i\hbar\Sigma\frac{da_n}{dt}\Psi_n = 0$$

Since the Ψ_n are solutions of Schrödinger's equation for the potential V_0, the expression in parentheses is zero. We now multiply the remaining terms by the complex conjugate of any one of the old wave functions, say Ψ_k, and integrate over all space:

$$\Sigma\, a_n\int_{-\infty}^{\infty}\Psi_k^* v\Psi_n\,dx = i\hbar\Sigma\frac{da_n}{dt}\int_{-\infty}^{\infty}\Psi_k^*\Psi_n\,dx$$

Recalling that

$$\int_{-\infty}^{\infty}\Psi_k^*\Psi_n\,dx = 0 \qquad \text{if } k \neq n$$

we have

$$\frac{da_k}{dt} = \frac{1}{i\hbar}\sum_n a_n\int_{-\infty}^{\infty}\Psi_k^*\, v\,\Psi_n\,dx$$

$$= \frac{1}{i\hbar}\sum_n a_n\int_{-\infty}^{\infty}\left(\psi_k^*\, e^{iE_kt/\hbar}\right) v \left(\psi_n\, e^{-iE_nt/\hbar}\right) dx$$

$$= \frac{1}{i\hbar}\sum_n a_n\, v_{kn}\, e^{-i(E_n-E_k)t/\hbar} \qquad (57)$$

where

$$v_{kn} = \int_{-\infty}^{\infty}\psi_k^*\, v\,\psi_n\,dx$$

The quantity v_{kn} is called the **matrix element** of the perturbation v taken between states n and k.

Equation 57 tells us how the value of one of the coefficients, a_k, changes with time. Now, if the perturbation v were zero, a_k would be a constant. So for very small perturbations we expect the rate of change of a_k to be small. As an approximation, we insert the initial values of the a_n in the right-hand side of Equation 57 and compute the time dependence of a_k as though the a_n on the right-hand side were constants. Since at $t = 0$ the system was in state m, then $a_m = 1$ and all other $a_n = 0$ initially. Consequently, we have for the value of a_k at time t

$$a_k = \frac{1}{i\hbar} \int_0^t v_{km} \, e^{-i(E_m - E_k)t/\hbar} \, dt \tag{58}$$

As shown in Section IV.B.5, the probability of finding the system in state Ψ_k is given by $|a_k|^2$. So the probability P that the system has changed to any other state is found by summing all the $|a_k|^2$ for which k is different from m:

$$P = \sum_{\substack{k \\ k \neq m}} |a_k|^2$$

In many cases of interest the perturbation v, once it has made its appearance, is constant in time. Then the matrix element v_{km} can be removed from the time integral of Equation 58 and a_k is seen to be proportional to v_{km}. Then the probability P that the system has made a transition to any other state is proportional to $|v_{km}|^2$.

7. Indistinguishable Particles

Large-scale objects can always be distinguished from one another by labeling them and continually observing their trajectories so that their positions are known at every moment. The situation for subatomic particles is different since not only are we unable to label such objects but, according to Heisenberg's uncertainty principle, we cannot observe their motion without significantly disturbing the system. Hence, if we have, for example, several electrons in the same system whose wave functions overlap, we will be unable to distinguish between them. This fact leads to some surprising results.

First it is necessary to find the correct wave function for a system containing two indistinguishable particles. Let us take the simple example in which two particles can only move along the x-axis and they do not interact with each other. Then the time-independent Schrödinger equation for this situation is

$$-\frac{\hbar^2}{2m}\frac{\partial^2 \psi(x_1,x_2)}{\partial x_1^2} - \frac{\hbar^2}{2m}\frac{\partial^2 \psi(x_1,x_2)}{\partial x_2^2}$$
$$+ U(x_1)\,\psi\,(x_1,x_2) + U(x_2)\,\psi\,(x_1,x_2)$$
$$= E\,\psi\,(x_1, x_2) \tag{59}$$

where x_1 and x_2 are the coordinates of particles 1 and 2 respectively (if they were known), $\psi\,(x_1, x_2)$ is the total wave function for the two particle system, $U(x_1)$ and $U(x_2)$ are the potential energies of the two particles, and E is the total energy of the system. Now, if two particles are truly indistinguishable and if it were possible to interchange them, there would be no way of knowing that the interchange had occurred. So $|\psi\,(x_1, x_2)|^2$ must remain the same if the coordinates of particles 1 and 2 are interchanged.

A reasonable guess for the form of $\psi\,(x_1, x_2)$ is that it is made up of some simple combination of the eigenfunctions for particles 1 and 2 when each is in the system alone. Calling ψ_{n_1} the nth eigenfunction of particle 1 with particle 2 removed and ψ_{m_2} the mth eigenfunction of particle 2 with particle 1 removed, it turns out that only two possibilities exist for $\psi\,(x_1, x_2)$, called the **symmetric** and **antisymmetric** forms:

$$\psi_{sym} = \frac{1}{\sqrt{2}} (\psi_{n_1} \psi_{m_2} + \psi_{m_1} \psi_{n_2})$$

$$\psi_{antisym} = \frac{1}{\sqrt{2}} (\psi_{n_1} \psi_{m_2} - \psi_{m_1} \psi_{n_2})$$

That either form satisfies Equation 59, and upon interchange of indexes 1 and 2 gives the same result for $|\psi|^2$, can be easily verified. We see that the so-called symmetric form does not change sign when particles 1 and 2 are interchanged while the antisymmetric form does change sign.

Whether a pair of like particles has the symmetric or antisymmetric wave function is physically significant. Suppose, for example, that a pair of particles has the antisymmetric form. Then, if the quantum numbers n and m are the same,

$$\psi_{antisym} = \frac{1}{\sqrt{2}} (\psi_{n_1} \psi_{n_2} - \psi_{n_1} \psi_{n_2}) = 0$$

This result, called the **Pauli exclusion principle,** says that two identical particles cannot both exist in the same quantum state if they are described by the antisymmetric wave function. Particles having the symmetric wave function, on the other hand, may occupy the same quantum state and in fact have an enhanced probability of doing so. Experiment shows that electrons, protons, neutrons, and other particles with half-integral spin (Section V.B) always have antisymmetric wave functions. Such particles are called **fermions**. Photons, tightly bound aggregates of an even number of fermions, and other particles having integral spin always have symmetric wave functions and are called **bosons**.

V. THE ATOMIC ELECTRONS

The atom consists of a central nucleus surrounded by orbiting electrons. Composed of protons and neutrons in approximately equal numbers, the nucleus contains almost the entire mass of the atom, yet occupies only 10^{-12} of the atomic volume. Protons have a positive electric charge of 1.60×10^{-19} C and a mass of 1.67×10^{-27} kg, which is 1836.1 times greater than the electronic mass. Neutrons have no electric charge and a mass slightly greater than that of the proton. The number of electrons in the atom is ordinarily equal to the number of protons and this is called the **atomic number** Z. Since the electric charge of the electron is equal in magnitude but opposite in sign to that of the proton, atoms are electrically neutral in their normal state.

It is primarily the Coulomb force between electrons and protons that holds the electrons in their orbits. The gravitational attraction is some 10^{39} times smaller in magnitude. Neutrons and protons are confined to the nuclear volume by the nuclear force as discussed in Section VI.B. Aside from the electrostatic attraction of electrons to the nucleus, there is little interaction between nucleus and electrons. The physics of the two systems can therefore be handled quite separately.

A. The Hydrogen Atom

As seen in Section IV.B.4 the proper way to treat systems of atomic dimensions is by the Schrödinger theory. We begin by writing the time-independent Schrödinger equation for the hydrogen atom, consisting of a proton at the origin of the coordinate system and an electron having coordinates x, y, and z. Since the proton is much more massive than the electron, it is approximately true that the proton remains stationary. Now, the force of attraction of the electron to the nucleus is the Coulomb force

$$F = \frac{1}{4\pi\epsilon_o} \frac{e^2}{r^2}$$

where e is the magnitude of the electric charge of the electron and proton and $r = \sqrt{x^2 + y^2 + z^2}$ is the distance from the electron to the nucleus. The potential energy corresponding to the Coulomb force is

$$U = -\frac{1}{4\pi\epsilon_o} \frac{e^2}{r}$$

where zero potential energy has been taken at infinity. The time-independent Schrödinger equation for this three-dimensional system is then

$$-\frac{\hbar^2}{2m}\left(\frac{\partial^2\psi}{\partial x^2} + \frac{\partial^2\psi}{\partial y^2} + \frac{\partial^2\psi}{\partial z^2}\right)$$
$$-\frac{1}{4\pi\epsilon_o} \frac{e^2}{\sqrt{x^2+y^2+z^2}} = E\psi \qquad (60)$$

It happens that solving Equation 60 is easier if it is written in terms of the spherical coordinates r, θ, and ϕ as shown in Figure 14. The relation between rectangular and spherical coordinates is

$$x = r \sin\theta \cos\phi$$

$$y = r \sin\theta \sin\phi$$

$$z = r \cos\theta$$

First the partial derivatives of Equation 60 must be transformed to spherical coordinates and then the transformed equation must be solved. Rather than perform these lengthy mathematical tasks (see Chapter 2, Section II) the results will be briefly presented.

1. The solutions ψ of Schrödinger's time-independent equation can be written as the product of three functions, each depending only on the single variable r, θ, or ϕ:

$$\psi_{n\ell m}(r, \theta, \phi) = R_{n\ell}(r)\, \Theta_{\ell m}(\theta)\, \Phi_m(\phi)$$

Here,

$$R_{n\ell}(r) = (\text{Polynomial in } r) \cdot e^{-\text{Constant}\cdot r}$$

$$\Theta_{\ell m}(\theta) = \text{Polynomial in } \cos\theta$$

$$\Phi_m = e^{im\phi}$$

Explicit formulas for R and Θ are given in Chapter 2, Section II.

2. The subscripts n, ℓ, and m of the eigenfunctions $\psi_{n\ell m}$ refer to the quantum numbers of the system. It is not surprising that three kinds of quantum numbers exist

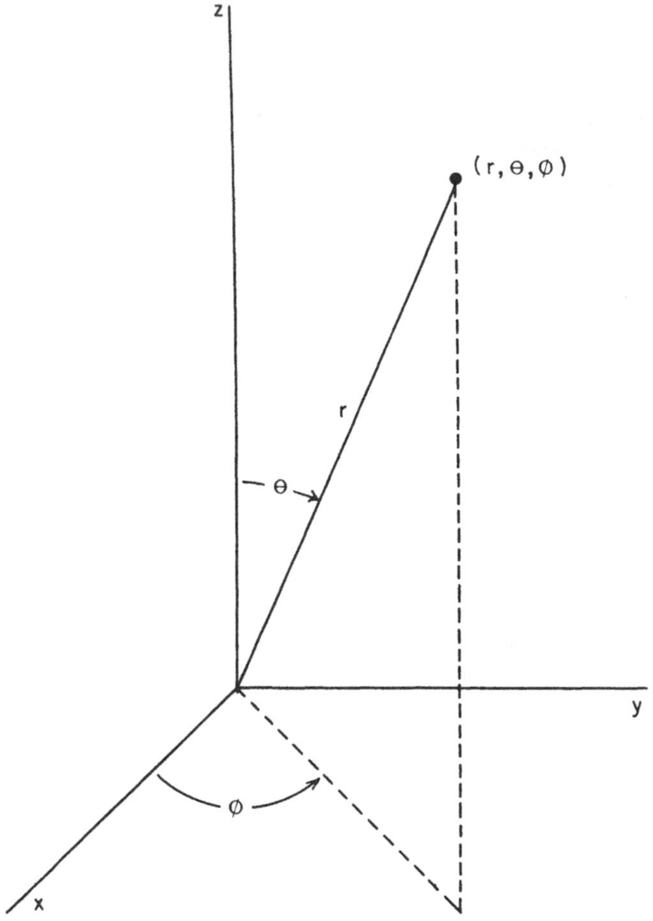

FIGURE 14. Spherical coordinates used in the mathematical description of the hydrogen atom.

for the hydrogen atom since it is a three-dimensional system. A fourth kind of quantum number having to do with a fourth degree of freedom, the electron spin, is discussed below. The values which the quantum numbers can assume are the following:

$$n = 1, 2, 3, \cdots$$

$$\ell = 0, 1, 2, \cdots, n - 1$$

$$m = -\ell, -\ell+1, \cdots, \ell$$

Thus the quantum number values are to some degree interrelated so that, for example, if $n = 3$, then ℓ can equal 0, 1, or 2. If $\ell = 2$, then m can have any of the values −2, −1, 0, 1, or 2.

3. As we have learned to expect for systems of extremely small dimensions, when the energy E of the system is less than the potential energy U then E takes on only discrete values. For the hydrogen atom the allowed energies of the electron are

$$E_n = -\frac{Me^4}{2\hbar^2 n^2}$$

where M is the mass of the electron. We see that the energy eigenvalues E_n depend only on the quantum number n whereas the eigenfunctions $\psi_{n\ell m}$ depend on ℓ and m as well.

4. The probability of finding the electron in a certain volume element dV of space will be

$$P = |\psi_{n\ell m}|^2 \, dV$$

For simplicity, let us ignore the angular dependence of $|\psi|^2$ and consider the probability of finding the electron at any angle between the radius r and r + dr. This is the region lying between two concentric spherical shells, one of radius r and the other of radius r + dr, and having the volume $dV = 4\pi r^2 dr$. Then the probability P_r of finding the electron between r and r + dr is

$$P_r = |R_{n\ell}(r)|^2 \, 4\pi r^2 dr$$

It turns out that P_r, which is plotted for several values of n and ℓ in Figure 7 of Chapter 2, depends mainly on n and only weakly on ℓ. Also, the larger the value of n, the larger is the radius at which P_r is a maximum. Thus, the quantum number n is closely associated with the distance from the nucleus at which the electron is likely to be located. The value of n is said to represent which shell the electron is in, implying that for each value of n there exists a rather restricted shell-like region about the nucleus in which the electron can be found. An electron having the quantum number n = 1 is said to be in the K shell, n = 2 corresponds to the L shell, etc.

$$n = 1, 2, 3, 4, \cdots$$

$$\text{shell} = K, L, M, N, \cdots$$

Although the value of ℓ does not appreciably affect the most probable distance from the nucleus at which the electron is likely to be found, it does affect the probability of finding the electron very near the nucleus. It happens that $R_{n\ell}(r)$ is proportional to r^{ℓ} for very small r so that P_r is proportional to $r^{2\ell+2}$. Thus, the chance of finding an electron near r = 0 is small unless ℓ is small. This fact becomes important in multielectron atoms where electrons spending time near the nucleus have a substantially reduced energy.

5. The entire angular dependence of $|\psi_{n\ell m}|^2$ dV is contained in the function $\Theta_{\ell m}$. This is because

$$|\Phi_m|^2 = e^{-im\phi} e^{im\phi} = 1$$

which is independent of angle. The nature of the angular dependence can be seen in Figure 8, Chapter 2. For electrons having $\ell = 0$, the wave functions have no angular dependence. As ℓ increases, the angular dependence becomes increasingly complicated. The probability of finding the electron along the z-axis is associated with the quantum number m: as the absolute value of m increases, the maximum probability shifts from the z-axis to the plane perpendicular to the z-axis.

6. It will be recalled (Section IV.B.5) that whenever an observable quantity Q that is written in operator form Q_{op} has eigenfunctions ϕ_i, then

$$Q_{op}\,\phi_i = q_i\,\phi_i$$

where the q_i (the eigenvalues of Q) are the only values that Q can have. It turns out that the energy eigenfunctions of the hydrogen atom, ψ_{nlm}, are eigenfunctions also of the square of the angular momentum L^2 and of the z-component of the angular momentum L_z:

$$L^2\,\psi_{n\ell m} = \ell(\ell + 1)\,\hbar\,\psi_{n\ell m}$$

and

$$L_z\,\psi_{n\ell m} = m\,\hbar\,\psi_{n\ell m}$$

These results give physical meaning to the quantum numbers ℓ and m. If a measurement is made on the angular momentum of an electron and on its z-components, the results will definitely be $\sqrt{\ell(\ell + 1)}\,\hbar$ and $m\,\hbar$, respectively. Thus, ℓ is associated with the magnitude of the angular momentum and m with its z-component.

For the isolated hydrogen atom, the choice of direction of the z-axis is arbitrary since there is no physical effect that indicates any preferred direction in space. So the component of angular momentum can be known definitely in any one direction. The ψ_{nlm}, however, are not eigenfunctions of L_x and L_y, implying that the values of these components are indefinite. This is the result of an uncertainty principle which states that no two angular momentum components can be known exactly at the same time.

B. Electron Spin

Precise measurements of the allowed energy levels of the hydrogen atom show that some levels actually consist of two closely spaced components. This **fine structure** is satisfactorily explained on the basis that the electron possesses an intrinsic angular momentum S in addition to its orbital angular momentum L about the nucleus. The intrinsic angular momentum S is termed **spin** and for purposes of visualization may be thought of as analogous to the rotation of the earth about its axis. Just as an extended charged body that is rotating has an associated magnetic moment (Section III.B.3) so the electron has a spin magnetic dipole moment. The splitting of energy levels is due to the fact that, from the point of view of the electron, the nucleus is orbiting about it. This orbiting electric charge gives rise to a magnetic field at the electron which interacts with the spin magnetic moment. The result is that the energy of the electron is slightly raised or lowered, depending on the direction of spin. This is the so-called **spin-orbit interaction** that occurs whenever the quantum number of the electron is not zero.

Because of this fourth degree of freedom of the electron, a fourth quantum number s is required. It happens that s can have only the single value

$$s = \tfrac{1}{2}$$

Analogously to the orbital angular momentum L, the magnitude of the spin angular momentum S is

$$|\mathbf{S}| = \sqrt{s(s+1)}\ \hbar = \frac{\sqrt{3}}{2}\ \hbar$$

Corresponding to the z-component of S is the quantum number m_s. Just as the quantum number m can range from $-\ell$ to $+\ell$ in integral steps, so m_s can range from $-s$ to $+s$:

$$m_s = -\tfrac{1}{2}, +\tfrac{1}{2}$$

The z-component of S can then have the values

$$S_z = m_s\,\hbar$$

Spin is incorporated in the eigenfunctions by multiplying ψ_{nlm} by a spin eigenfunction σ that can have only two values corresponding to the two possible values of m_s. The complete eigenfunction is therefore designated by four subscripts:

$$\psi_{n\ell m_\ell}\sigma_{m_s}$$

where the symbol m_l replaces the previously used m to make the notation more consistent. It is not necessary to include the quantum number s as a subscript since it has only one value.

The inclusion of σ_{m_s} in the eigenfunction of the electron has certain implications with regard to systems containing two or more electrons. It will be recalled (Section IV.B.7) that the nature of electrons is such that when two are interchanged in a system, the total eigenfunction changes sign (is antisymmetric). With the inclusion of the spin eigenfunction, it can be shown that four antisymmetric eigenfunctions are possible for a two electron system:

$$\psi = \begin{cases} \frac{1}{\sqrt{2}}\left[\psi_a(1)\,\psi_b(2) + \psi_b(1)\,\psi_a(2)\right]\ \frac{1}{\sqrt{2}}\left[\sigma_{1/2}(1)\,\sigma_{-1/2}(2) - \sigma_{-1/2}(1)\,\sigma_{1/2}(2)\right] \\ \frac{1}{\sqrt{2}}\left[\psi_a(1)\,\psi_b(2) - \psi_b(1)\,\psi_a(2)\right]\ \sigma_{1/2}(1)\,\sigma_{1/2}(2) \\ \frac{1}{\sqrt{2}}\left[\psi_a(1)\,\psi_b(2) - \psi_b(1)\,\psi_a(2)\right]\ \frac{1}{\sqrt{2}}\left[\sigma_{1/2}(1)\,\sigma_{-1/2}(2) + \sigma_{-1/2}(1)\,\sigma_{1/2}(2)\right] \\ \frac{1}{\sqrt{2}}\left[\psi_a(1)\,\psi_b(2) - \psi_b(1)\,\psi_a(2)\right]\ \sigma_{-1/2}(1)\,\sigma_{-1/2}(2) \end{cases}$$

where the 1 or 2 in parentheses refers to electrons 1 or 2, a and b are the spatial quantum numbers, and ½ and −½ are the two possible projections of the spin on the z-axis. It is seen that the first of these eigenfunctions is composed of a symmetric combination of spatial eigenfunctions multiplied by an antisymmetric combination of spin eigenfunctions. It is necessary that the symmetry of the space and spin components be opposite so that the total eigenfunction is antisymmetric. The first eigenfunction represents the situation in which the two electrons have their spins oriented in opposite directions which is termed the **singlet state.** The remaining three eigenfunctions have an antisymmetric spatial part and a symmetric spin part. These represent the so-called **triplet** states in which the electrons have their spins oriented parallel to each other but in three different directions with respect to the z-axis. Electrons in the triplet state are unlikely to be found close together since then 1 and 2 represent the same location so that the spatial part of their eigenfunction vanishes. In contrast, electrons in the singlet state are likely to be found close together.

Because of the spin-orbit interaction, the quantum numbers m_l and m_s are not the best to use. The reason is that since the motion of the electron in its orbit is affected by the electron spin, the orbital angular momentum **L** and the spin angular momentum **S** both carry out a precessional movement about the vector **J** which is the vector sum of the two:

$$\mathbf{J} = \mathbf{L} + \mathbf{S} \tag{61}$$

The precession does not affect the magnitude of **L** and **S** but it does cause L_z and S_z to vary in time. Consequently, m_l and m_s no longer have definite values and may not be used to specify the eigenfunctions of the atom.

The total angular momentum **J**, however, does have a fixed z-component and provides a satisfactory replacement for L_z and S_z. As for all angular momentum vectors in the quantum theory, the magnitude and z-components of **J** are given by the relations

$$|\mathbf{J}| = \sqrt{j(j+1)}\,\hbar$$

$$J_z = m_j\,\hbar$$

where j and m_j are the quantum numbers associated with **J**. Because of Equation 61, we expect the quantum number j to be related to ℓ and s, and since s has only the single value ½, it happens that j is permitted only the two values

$$j = \ell - ½,\ \ell + ½$$

If $\ell = 0$, then $j = ½$ only. Because ℓ always has integral values, j will always have half-integral values such as 1/2, 3/2, 5/2, $\cdots$. The z-component of **J** can have any of the values

$$m_j = -j, -j+1, \cdots, +j$$

The four quantum numbers required to completely specify the eigenfunctions of the hydrogen atom then are n, ℓ, j, and m_j.

C. Many-Electron Atoms

The physics of atoms having two or more electrons is considerably more complicated than that of the hydrogen atom because of the complex interactions among the electrons. Fortunately, by making several simplifying assumptions it is possible to treat the multielectron atom in much the same way as the hydrogen atom. The basis of this approach is the Hartree theory, in which it is assumed that the interactions among electrons average out in a rather simple way. The idea is that from the standpoint of any one electron, the presence of the others is only to make it appear as though the amount of positive charge in the nucleus varies according to the distance of the electron from the nucleus. For example, in an atom with Z electrons and protons, an outermost electron "sees" (Z−1) electrons between it and the nucleus so that the nuclear charge appears to be +e. On the other hand, an innermost electron is not shielded from the nucleus and experiences a nuclear charge of $+Ze$. Mathematically, the effect of shielding is expressed by using an effective nuclear charge $Z'e$ which depends on the distance r of the electron from the nucleus. The potential energy of any electron is then

$$U = -\frac{1}{4\pi\epsilon_o}\frac{e^2 Z'(r)}{r}$$

Hence, the effect of having many electrons present in the atom is to alter the simple $1/r$ dependence of the potential energy, causing its magnitude to decrease faster with r than in the single-electron atom. This exaggerates the dependence of the total energy of the electron on the distance of the electron from the nucleus.

Because in this approximation the form of the potential energy for each electron is so similar to that for the hydrogen atom, the wave function for each electron is similar to a hydrogen atom wave function. The total wave function of the atom is then given as the product of the individual electron wave functions. Ignoring for the moment any spin-orbit interactions, we specify the state of each electron by the four quantum numbers n, ℓ, m_l, and m_s.

As was seen for the hydrogen atom, the smaller the value of the quantum number ℓ, the greater is the probability of finding the electron very near the nucleus. This fact has no effect on the total energy E of the electron in the hydrogen atom. However, for the multielectron atom with its enhanced dependence of potential energy on r, the total energy of electrons having a small value of ℓ is less than that for electrons having large ℓ values. In fact for atoms with many electrons the ℓ-dependence of E can be greater than the n-dependence. Because of the significance of the quantum number ℓ in multielectron atoms, a terminology has arisen in which an electron in a certain shell specified by n is in addition said to be in a **subshell** specified by ℓ. The relation between the ℓ-value and the subshell designation is as follows:

$$\ell = 0, 1, 2, 3, 4, 5, \cdots$$

$$\text{subshell} = s, p, d, f, g, h, \cdots$$

Electrons in atoms will occupy those states which minimize the energy of the atom since, as described below (Section V.D), if an electron can move to a lower energy level it will do so with the emission of radiation. As we have seen, it is the quantum numbers n, ℓ, m_l, and m_s which specify the electronic state. Of these, only n and ℓ appreciably affect the energy of the electron, the energy being lowest when n and ℓ are lowest. Now, were it not for the Pauli exclusion principle, all electrons would reside in the lowest energy state with $n = 1$, $\ell = 0$. However, since no two electrons in an atom can have exactly the same quantum numbers, electrons are forced to take on successively larger n and ℓ values for atoms having greater and greater numbers of electrons. To see how this works, we recall the possible values of the quantum numbers for each electron:

$$n = 1, 2, 3, \cdots$$

$$\ell = 0, 1, 2, \cdots, n-1$$

$$m_\ell = -\ell, -\ell+1, \cdots, +\ell$$

$$m_s = -\tfrac{1}{2}, +\tfrac{1}{2}$$

The hydrogen atom in its ground state (state of lowest energy) has an electron in the state $n = 1$, $\ell = 0$, $m_\ell = 0$, $m_s = +½$. Helium has a second electron which occupies the state $n = 1$, $\ell = 0$, $m_\ell = 0$, and $m_s = -½$. At this point all possible $n = 1$ states are filled. Hence, lithium with three electrons has two electrons in $n = 1$ states and the third in the state $n = 2$, $\ell = 0$, $m_\ell = 0$, and $m_s = +½$. Because of its larger value of n and larger mean distance from the nucleus, the third electron has considerably higher energy. As more electrons are added, succeeding shells and subshells fill. The order in which these shells and subshells fill depends on the exact n- and ℓ-dependence of the total energy E. Since it is the electrons farthest from the nucleus that most readily participate in molecular combinations, the chemical properties of the elements are principally determined by the energy states of the electrons in the outermost shells and subshells.

It was seen in the case of the hydrogen atom that because of the spin-orbit interaction the orbital angular momentum **L** and the spin angular momentum **S** of the electron precess about the vector **J** that is their sum. Now, in a many-electron atom one might expect this phenomenon to also occur. Furthermore, it is conceivable that because electrons do interact somewhat with each other that these interactions cause the individual angular momentum vectors J to couple so that they precess about a total vector J′ which is their sum:

$$\mathbf{J}' = \mathbf{J}_1 + \mathbf{J}_2 + \cdots$$

Indeed, for atoms of large atomic number this does occur and is called **JJ coupling.** A similar effect is seen in the protons and neutrons of the nucleus.

In more advanced treatments of quantum mechanics the rules of angular momentum addition are developed. Briefly, these rules state that if

$$\mathbf{J} = \mathbf{j}_1 + \mathbf{j}_2$$

then the angular momentum quantum number J associated with **J** can take on any of the values

$$J = |j_1 - j_2|, |j_1 - j_2| + 1, \cdots, j_1 + j_2$$

where j_1 and j_2 are the quantum numbers associated with the vectors $\mathbf{j}_1$ and $\mathbf{j}_2$. A special case which is often of importance is that in which two identical fermions have the same n, ℓ, and j quantum numbers. Then the Pauli exclusion principle requires that J can only have the values

$$J = 0, 2, 4, \cdots, 2j-1$$

In general, if the z-component of **J** is M, then

$$M = m_1 + m_2$$

where m_1 and m_2 are the z-components of $\mathbf{j}_1$ and $\mathbf{j}_2$.

For atoms of low and intermediate Z, it is much more common for the individual orbital angular momenta $\mathbf{L}_1, \mathbf{L}_2, \cdots$, to precess about their sum **L′** and for the individual spin angular momenta $\mathbf{S}_1, \mathbf{S}_2, \cdots$, to precess about their sum **S′**. Then a spin-orbit interaction between **L′** and **S′** causes these to couple, forming a total angular momentum **J′**:

$$\mathbf{J}' = \mathbf{L}' + \mathbf{S}'$$

Termed LS or **Russell-Saunders coupling**, this phenomenon occurs when the effect of electric repulsion among electrons is large relative to the spin-orbit interaction. For example, it can be shown that the average distance between two electrons tends to be greatest when their individual L vectors are parallel and when their S vectors are parallel. In this state the electrons experience the least Coulomb repulsion and the energy of the atom is minimized. Except in atoms of large Z, energy effects of this kind are substantially greater than those due to the spin-orbit interaction.

D. Characteristic X-Rays and the Auger Effect

The mechanism whereby atoms absorb and emit energy will now be considered. In the absence of any disturbance, an atom exists in a state of definite energy and, left undisturbed, will forever remain in that state. For the atom to emit or absorb energy requires that its wave function be altered and this is possible only when a perturbation is applied to the atom, as discussed in Section IV.B.6. In practice, excitation of the atom may result from collisions with other atoms or from photon or charged particle bombardment. In the excitation process one or more atomic electrons are moved to states of higher energy. The atom subsequently loses its excitation energy with the electrons returning to their lowest energy levels by emitting electromagnetic radiation (photon emission), by ejecting an electron from the atom (the Auger effect), by transferring its excitation energy to another atom in a collision (inelastic collisions), or by other means.

Of special interest is the event in which an atom loses its excitation energy by the emission of photons. In the process called **spontaneous emission**, photons are produced even though all externally applied perturbations may have been removed from the excited atom. At first glance, spontaneous emission would seem to contradict the idea that a perturbation is necessary to alter atomic energy states. It happens, however, that an electromagnetic radiation field exists even in the absence of an externally applied photon beam and that this field, even when "empty" of photons, is capable of interacting with the electrons of the atom. Through its interaction with the radiation field, an electron in an excited state can drop to a lower energy state with the production of a photon. The energy of the photon is equal to the difference in energy of the atom before and after the transition. It is also possible for atoms to be induced to emit photons by an externally applied field in the process of **stimulated emission**. This forms the basis of operation of the laser. Whether it is atomic excitation or de-excitation in their various forms, the probability of occurrence is proportional to the square of the matrix element of the particular perturbation taken between the initial and final states of the atom, as discussed in Section IV.B.6.

In terms of individual electrons, the excitation and de-excitation of atoms correspond to shifts of one or more electrons between atomic shells. Of course, electron transitions must not violate the Pauli exclusion principle. For example, in excitation, an electron must move to an unfilled shell of higher energy or leave the atom entirely. The vacancy thus produced can then be filled by an electron of higher energy. Transitions of electrons between outer shells involve relatively small energies and the resulting electromagnetic radiation is termed the **optical spectrum** of the atom since it may be visible to the eye.

The excitation of inner shell electrons, on the other hand, requires a comparatively large energy because of their strong attraction to the protons of the nucleus. The filling of an inner shell vacancy by a less tightly bound electron has the effect of moving the vacancy outwards, and several electron transitions may occur before the vacancy reaches the outermost shell to be filled by an electron from outside the atom. The energetic photons which are concomitantly produced are termed **characteristic X-rays.** As an example of characteristic X-ray production, we consider an atom which has had a K-shell electron removed from it. The energy of the atom is increased above its ground state energy by an amount B_k, termed the binding energy of the K shell electron. If an L shell electron were then to fill the K shell vacancy, the atom would be missing an electron from the L shell and would have an excitation energy of B_L. The energy of the emitted X-ray, E_{Xray}, would then be

$$E_{Xray} = B_K - B_L$$

Since the binding energies of the inner shell electrons increase smoothly with increasing atomic number Z of the nucleus (approximately as $(Z-1)^2$), X-ray energies also vary regularly with Z and are "characteristic" of the element from which they are emitted.

When an electron makes a transition to a lower energy level, an alternative to X-ray emission is the **Auger effect.** Here, the energy of de-excitation, instead of appearing in the form of a photon, is used to eject one of the electrons from the atom. A second vacancy is then present in the atom which must be filled. As these vacancies move toward the outermost shells, additional photons or ejected electrons are produced. The energy of the ejected (or Auger) electron is equal to the energy of the atom before the original vacancy was filled minus the energy of the atom after the Auger electron has departed. To an approximation, the Auger electron energy E_{Auger} may be computed as the energy which the X-ray would have had, E_{Xray}, minus the binding energy of the Auger electron B_{Auger}:

$$E_{Auger} = E_{Xray} - B_{Auger}$$

The probability of the Auger effect relative to photon emission is large in elements of low atomic number.

VI. THE ATOMIC NUCLEUS

At the center of the atom resides the nucleus, a dense assembly of protons and neutrons (collectively, **nucleons**). Four invariant properties of nucleons are their mass at rest, electric charge, intrinsic spin, and magnetic dipole moment. The mass of the proton is 1.67×10^{-27} kg while that of the neutron is about 0.1% greater. The principal difference between proton and neutron is the electric charge (which for the proton is equal and opposite to that of the electron), 1.60×10^{-19} C for the proton and zero for the neutron. Like the electron, nucleons possess an intrinsic angular momentum with spin quantum number ½. Because of their half-integral spin, these particles all obey the Pauli exclusion principle. In classical physics a spinning electric charge is expected to have a magnetic dipole moment. It is therefore not surprising that the proton possesses a magnetic moment, 1.41×10^{-26} J-m²/wb, although this value is about three times larger than expected. Remarkably, the neutron also has a magnetic moment, -9.65×10^{-27} J-m²/wb, where the minus sign means that the magnetic moment points in a direction opposite to the spin angular momentum vector. A non-zero magnetic moment for the neutron indicates that its overall neutral charge consists of a nonhomogeneous distribution of positive and negative charge.

A. Nomenclature

The number of protons in the nucleus is referred to as the **atomic number** Z and the number of neutrons as the **neutron number** N. The total number of nucleons Z + N is termed the **mass number** A of the nucleus. A substance composed of atoms all having the same atomic number is said to be one of the chemical elements, more than 90 of which exist in nature. The composition of a nucleus is shown symbolically in the fashion $^{A}_{Z}X$ where X represents the chemical symbol corresponding to the atomic number Z and the superscript A is the mass number. Often the subscript Z is omitted since it and the chemical symbol provide the same information. Thus, $^{19}_{9}F$ represents a nucleus having 9 protons (the element fluorine), 19 nucleons, and hence 10 neutrons.

Any particular nuclear species characterized by its number of protons Z and neutrons N is termed a **nuclide**. Different nuclides having the same Z are called **isotopes** of the same element, such as the carbon isotopes $^{12}_{6}C$, $^{13}_{6}C$, and $^{14}_{6}C$. The terms **radionuclide** and **radioisotope** refer to nuclides or isotopes which emit radiation. (The widespread use of the word "isotope" to mean "nuclide" or "radionuclide" is falling into disfavor.) **Isobars** are different nuclides having the same A, e.g., $^{12}_{6}C$ and $^{12}_{5}B$. **Isotones** are different nuclides having the same N, e.g., $^{18}_{8}O$ and $^{19}_{9}F$. Nuclear **isomers** are different energy states of the same nuclide in which the nucleus can exist for a relatively long time. To distinguish symbolically between isomeric states, the superscript m for "metastable" is included with the mass number when referring to an excited isomeric state. For example, ^{99m}Tc refers to a long-lived excited state of the nuclide ^{99}Tc.

B. Nuclear Properties and Systematics

Simply pictured, the nucleus is an approximately spherical assembly of small incompressible objects. Its volume is proportional to the number of nucleons A so that its radius R is proportional to the cube root of A:

$$R = 1.1\ A^{1/3}\ \text{fm}$$

where the fermi (fm) is equal to 1×10^{-15} m. A typical nucleus with A = 100 has then a radius of about 5 fm as compared with an atomic radius of 10^5 fm. Supporting the idea of the nucleus as a collection of incompressible objects, the nuclear density does not vary much throughout the nuclear interior nor from nuclide to nuclide, having a value of about 2×10^{17} kg/m^3.

It cannot be electromagnetic or gravitational forces that account for the extraordinarily high density of the nucleus. Protons mutually repel each other because of their like charge, and the gravitational attraction among nucleons is much too weak to counteract this effect. Instead, it is the so-called **strong force** (as distinguished from the weak interaction associated with beta decay) which provides the attraction necessary to overcome the Coulomb repulsion among protons. The nuclear force is complicated and incompletely understood but some of its principal features are well established. The attraction between two nucleons has a short range, falling to zero when their separation is greater than about 2 fm. However, nucleons approaching closer than about 0.4 fm experience a strongly repulsive force. It is in part this repulsion which prevents the nucleus from collapsing. Nucleons do not interact strongly with every other nucleon in the nucleus or even with every nucleon within the short range of the nuclear force. Instead, nuclear forces **saturate** so that a nucleon is strongly bound to no more than about three others. Whether acting between protons or neutrons, the nuclear force is approximately the same.

One might expect that the mass of a nucleus could be easily calculated by summing the individual masses of its protons and neutrons. However, as seen in Section IV.A.7 the mass and energy of a system are interrelated. Because of the attractive nuclear force, nucleons have a lower energy (hence, lower mass) when they are together than when they are apart. The **binding energy** of the nucleus is defined as the energy required to dissociate it into widely separated nucleons. From the mass-energy relationship of Equation 33, the binding energy B may be computed as the difference between the sum of the masses of the separate nucleons and the measured mass of the nucleus, multiplied by c^2:

$$B = (Z\,M_p + N\,M_n - {}^{A}_{Z}M)\,c^2$$

where M_p, M_n, and ${}^{A}_{Z}M$ are the masses of the proton, neutron, and nucleus, respectively. The mass that is lost when a number of nucleons are assembled to form a nucleus is typically 1% of the total.

Rather than express nuclear and atomic masses in kilograms, a scale is used in which the mass of the ^{12}C atom is taken to be exactly 12 mass units (u). On this scale the mass of the proton, neutron, and electron are

$$M_p = 1.00728 \text{ u}$$

$$M_n = 1.00867 \text{ u}$$

$$M_e = 0.00055 \text{ u}$$

Alternatively, masses may be expressed in energy units by use of Equation 33. The commonly used unit of energy in atomic and nuclear physics is the **electron volt** (eV). One electron volt is the kinetic energy an electron acquires when accelerated in the electric field produced by a difference of potential of one volt and is equal to 1.60×10^{-19} J. Multiples of the electron volt are the keV (10^3 eV) and MeV (10^6 eV). Expressed in energy units, the masses of the proton, neutron, and electron are

$$M_p = 938.256 \text{ MeV}$$

$$M_n = 939.550 \text{ MeV}$$

$$M_e = 0.511 \text{ MeV}$$

From the measurement of nuclear masses it is found that the nuclear binding energy is approximately proportional to the number of nucleons A, as shown in Figure 15. The proportionality of binding energy to mass number is consistent with the idea that a nucleon is bound to only a limited number of neighboring nucleons. The fact that the binding energy per nucleon decreases at low values of A is a result of these nuclides having a greater proportion of nucleons at the nuclear surface where they are not as effectively bound. At large values of A there is a loss of binding energy due to Coulomb repulsion among protons. From these simple considerations it is possible to fairly accurately fit the binding energies of the stable nuclides as a function of A with a formula having only several terms:

$$B = a_1\,A - a_2\,A^{2/3} - a_3\,\frac{Z^2}{A^{1/3}}$$

where a_1, a_2, and a_3 are empirically determined constants. The first term, proportional to A or the nuclear volume, is the binding energy which the nucleus would have if

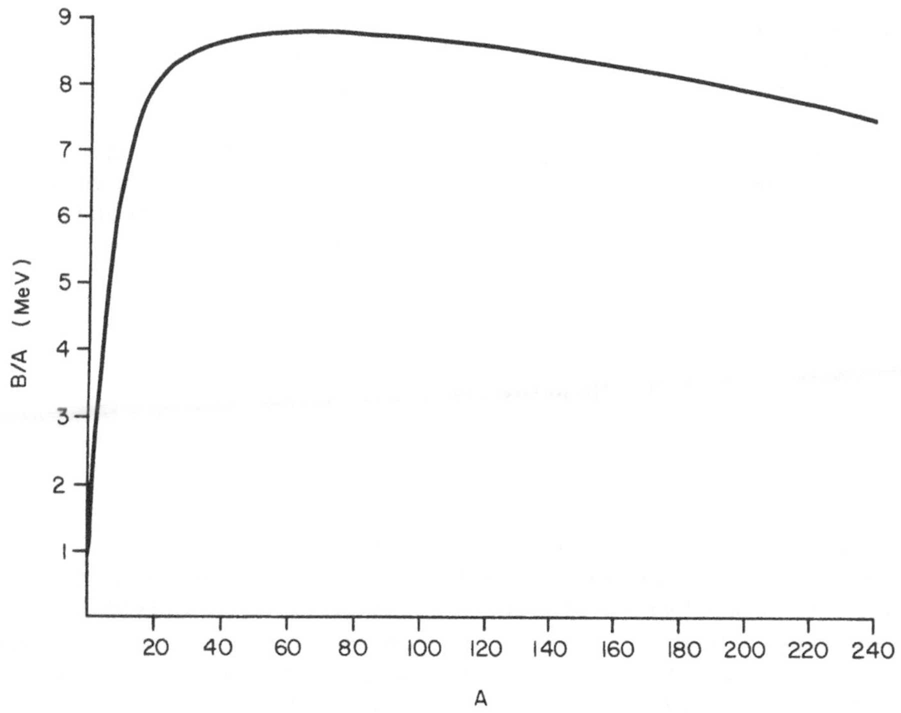

FIGURE 15. Binding energy per nucleon of stable nuclides as a function of mass number A.

every nucleon were in the interior where the forces on it are saturated. The fact that nucleons at the surface experience less binding because they are not completely surrounded by nucleons is expressed by the second term, which is proportional to the surface area $A^{2/3}$ of the nucleus. The third term, which is proportional to the square of the charge Z of the nucleus divided by the nuclear radius $A^{1/3}$, represents the disruptive Coulomb energy. Because a drop of ordinary liquid is similar in some respects to nuclear matter, also having a total binding energy (or heat of vaporization) that is proportional to the volume of the liquid and having surface tension effects, a visualization of the nucleus in these terms is called the **liquid drop model**.

Stable nuclides have large binding energies compared to those of their neighboring radionuclides. A study of the stable nuclides reveals several regularities related to their Z and N values which reflect on the nature of nuclear forces. First, N is approximately equal to Z for stable nuclides, although as A increases the number of neutrons predominates over protons. Second, nuclei having too many nucleons are unstable, there being no stable nuclides with A greater than 209. Coulomb repulsion among protons is responsible for the inherent instability of heavy nuclei and the predominance of N over Z. This force becomes increasingly important for large nuclei because of its long range compared to that of the nuclear force. Third, nuclides having even Z and even N are more stable than odd Z or N nuclides. For example, only four odd-Z-odd-N nuclides are nonradioactive while the number of stable even-Z-even-N nuclides is very large. Fourth, the binding energy of nuclides is particularly great when their Z or N is equal to any of the numbers 2, 8, 20, 28, 50, 82, or 126. The existence of these "magic numbers" indicates that there are shell effects for nucleons similar to those in atomic physics.

C. Radioactive Decay

Nuclear decay or **disintegration** is a process whereby a nucleus spontaneously lowers its energy with the emission of radiation. This may be accomplished by a nucleon changing its identity inside the nucleus or by the production of electromagnetic or particulate radiation. In order to conserve mass and energy in nuclear decay, it is necessary that the mass of the atom which is to decay (the **parent** atom) be equal to the mass of the atom after decay (the **daughter** atom) plus the rest mass of any radiation emitted plus the kinetic energy T of the radiation and of the daughter atom:

$$M_{Parent}\, c^2 = M_{Daughter}\, c^2 + M_{Radiation}\, c^2 + T$$

Spontaneous decay cannot occur unless

$$M_{Parent} \geqq M_{Daughter} + M_{Radiation}$$

since otherwise the kinetic energy of the products would be negative.

In nuclear decay several particles arise which we do not otherwise encounter. Some of these are associated with the fact that all fundamental constituents of the universe have a so-called **antiparticle**. A particle and its antiparticle are identical in all respects except that their electric charges are of opposite sign. For example, the antiparticle of the electron is the **positron**, having the same mass and spin but a positive charge. When a particle and antiparticle meet, their masses are transformed into radiative energy in an event called **annihilation**. The annihilation of an electron-positron pair produces two photons of 511 keV each (the energy equivalent of an electron mass) which are emitted in opposite directions. Another particle-antiparticle pair are the **neutrino** and **antineutrino** encountered in beta decay. These particles, which have little or no mass and no electric charge, interact so weakly with matter that they are virtually undetectable. The **antiproton** and **antineutron**, because of their large masses, occur only as the result of high-energy nuclear reactions. The important processes by which nuclei decay are discussed below.

1. Gamma Ray Emission and Internal Conversion

As a result of external bombardment or of a nuclear decay process, a nucleus may find itself in an excited state. The excited nucleus may then make a transition to a lower energy level by the emission of a photon. This process is analogous to the emission of characteristic X-rays when atomic electrons move to lower energy states. Photons emitted as a result of nuclear de-excitation are termed **gamma rays** and are no different in character from other electromagnetic radiation. The energy of the gamma radiation is the difference in energy between the nuclear states.

As an alternative to gamma ray emission, the energy of de-excitation may be used to eject an atomic electron in a process called **internal conversion**. Internal conversion is analogous to the Auger effect in atomic physics. The energy of the internal conversion electron is equal to the energy which the gamma ray would have had minus the binding energy of the electron. Following internal conversion, a vacancy is left in an atomic shell which, when filled, results in the production of one or more characteristic X-rays or Auger electrons.

No change in Z or N results from gamma ray emission or internal conversion.

2. Beta-Minus Decay

The processes of beta-minus, beta-plus, and electron capture decay are inclusively referred to as **beta decay**. From a theoretical point of view they are very similar, being the result of a force, the **weak interaction**, between neutrons and protons, electrons and neutrinos.

Nuclides that have an excess of neutrons can decrease their energy in a process called **beta-minus** or **negatron decay**. In the nucleus a neutron changes into a proton so that the neutron number N decreases by one while Z increases by one, leaving A unaltered. Carrying away the de-excitation energy, an electron e^- and antineutrino $\bar{\nu}$ are created in the nucleus and ejected from it:

$$n \rightarrow p + e^- + \bar{\nu}$$

The ejected electron is referred to as a **beta particle**.

Typically, the nuclide which is to beta-minus decay is in its ground state. The decay generally goes to an excited state of the daughter nuclide, after which several gamma rays or internal conversion electron emissions occur before the ground state of the daughter is reached.

3. Beta-Plus Decay

Beta-plus or **positron decay** is the "mirror image" of beta-minus decay: in the nucleus a proton changes into a neutron, and a positron e^+ and neutrino ν are emitted from the nucleus

$$p \rightarrow n + e^+ + \nu$$

The effect is to reduce Z by one, increase N by one, and to leave A unaltered. When the positron subsequently encounters an electron, the two annihilate with the production of the annihilation quanta. From a straightforward consideration of mass-energy conservation (Volume II, Chapter 2), it is found that beta-plus decay cannot occur unless the energy which the parent atom would lose in the process is greater than the energy equivalent of two electron rest masses (1022 keV).

4. Electron Capture

An alternative to beta-plus decay is **electron capture**. An atomic electron is captured by the nucleus, usually from the K shell, and combined with a proton to form a neutron. Carrying away the energy lost in the decay, a neutrino is emitted from the nucleus:

$$p + e^- \rightarrow n + \nu$$

As with beta-plus decay, the neutron number N is increased by one, Z is decreased by one, and A remains the same. Following the capture of the electron, there is a vacancy in the shell from which it was captured, and characteristic X-rays or Auger electrons are produced when this vacancy is filled.

5. Alpha Decay

A high-Z nuclide experiences strong forces of Coulomb repulsion which could be reduced if the nucleus were able to rid itself of some of its protons. For this event to be energetically possible, it is necessary that the mass of the parent atom be greater than the sum of the masses of the daughter atom and of any nucleons emitted. Now, an assembly of two protons and two neutrons, the ^4_2He nucleus or **alpha particle**, has a particularly large binding energy so that its mass is comparatively small. In fact, of all nuclear systems that could be emitted, only the ejection of an alpha particle is energetically possible. In alpha decay, Z and N decrease by two and A decreases by four:

$$^{A}_{Z}M \rightarrow {}^{A-4}_{Z-2}M + {}^{4}_{2}He$$

Alpha decay is hindered by the fact that inside the nucleus the nucleons are trapped in a potential well. In fact, from the standpoint of classical physics an alpha particle does not have sufficient energy to escape and alpha decay can never occur. However, as seen in Example 15, for systems of very small dimensions there is a finite probability that a particle may tunnel through a potential barrier that is classically impenetrable. Once beyond the nuclear radius where the nuclear force falls to zero, the alpha particle is accelerated away by Coulomb repulsion and the decay ensues.

D. Shell Model of the Nucleus

At first glance, one would expect nucleons and atomic electrons to behave quite differently. Not only do nucleons experience an entirely different force but their environment is unlike that of the electrons. Nucleons are not attracted to a single central particle and their close packing would suggest frequent collisions accompanied by continually changing energies and angular momenta. Nevertheless, for several reasons nuclear and atomic physics are remarkably similar. First, basic principles that applied to electrons, such as the Schrödinger theory, also apply to nucleons. Second, nucleons and electrons alike are fermions, obeying the Pauli exclusion principle. Third, although the nuclear force has little in common with the Coulomb force felt by electrons, the net effect of the nuclear force on nucleons is in many ways similar to the average force felt by electrons in a multielectron atom. The reason is that, surprisingly, mutual interactions among nucleons are comparatively weak so that to a rough approximation each nucleon moves independently of the others. It is the Pauli exclusion principle that restricts the interactions among nucleons: in a collision, unless a nucleon is able to move to an energy state which is unoccupied by other nucleons, it cannot change its state at all.

Because of the constant density of the nucleus throughout its interior, an inner nucleon is attracted equally in all directions and so experiences no net force. However, at the nuclear surface, which we assume to be spherical, nucleons feel a strong attraction toward the nuclear center. The corresponding potential energy is approximately constant within the nucleus, rising sharply at the nuclear radius R and becoming constant again outside the nucleus. Several reasonably simple mathematical approximations to this desired shape exist, e.g., the square well potential of Example 15 extended to three dimensions. When such a potential energy function is substituted into the time-independent Schrödinger equation, solutions of the equation yield a set of permissible energy levels which are specified by the four quantum numbers n, ℓ, m_ℓ, and m_s. Since protons repel each other, their potential energies and allowed energy states are somewhat higher than those of neutrons. The Pauli exclusion principle acts separately on protons and neutrons to fill energy levels so that no two identical nucleons occupy exactly the same quantum state. When energy levels have been filled so as to produce the lowest possible energy for the nucleus, the nucleus is said to be in its ground state.

An additional fact is needed to obtain correspondence with experimental data regarding nuclear energy levels. This fact is that the nuclear force, for reasons that are not well understood, causes the orbital motion of each nucleon to interact with its intrinsic spin. The effect of this spin-orbit interaction is much greater for nucleons than for electrons, resulting in significant shifts in nuclear energy levels. As shown in detail in Volume II, Chapter 1, the use of a reasonable potential energy function along with a spin-orbit interaction produces a set of energy levels whose order and spacing are in quite satisfactory agreement with experimental findings.

The picture of the nucleus as a collection of particles each experiencing a spin-orbit interaction and each moving more or less independently in a spherically symmetric potential well is termed the **shell model** of the nucleus. It provides plausible explanations for many nuclear properties. For example, the angular momenta of nuclear ground states are successfully predicted in nearly all cases. Also compatible with the shell model is the fact that nuclei are especially stable if their Z or N is equal to 2, 8, 20, 28, 50, 82, or 126. These so-called "magic numbers" occur whenever there is a large energy gap between successive energy levels. In atomic physics the analogous situation is the stability associated with the noble gases. Large energy gaps at the magic numbers are correctly predicted by the shell model when a spin-orbit interaction of proper strength is included.

As with electrons, the spin-orbit interaction causes the spin angular momentum s and the orbital angular momentum ℓ of each nucleon to precess about their vector sum j so that the quantum numbers appropriate for identifying different states are n, l, j. and m_j. The individual j-vectors in turn add together to form a total angular momentum vector for the nucleus (jj coupling). The manner in which individual angular momenta add is particularly simple for nuclides in their ground state: when there is an even number of protons (or neutrons), their angular momenta add in such a way as to give zero total angular momentum. For an odd number of protons or neutrons, the total angular momentum will be that of the odd nucleon in the highest-occupied energy level.

An explanation for the manner in which nucleons add their angular momenta is that identical nucleons (protons or neutrons) in the same energy level cancel their angular momenta in pairs. Presumably, such a phenomenon could occur if there existed in addition to the average force on each nucleon a smaller residual force which binds pairs of nucleons when their angular momenta point in opposite directions. Evidence for this **pairing energy** is found in the fact that even-Z-even-N nuclides have generally larger binding energies than expected. The effect of the pairing interaction is to cause even-Z-even-N nuclides to be relatively inert since comparatively large amounts of energy are required to excite these nuclei and because their zero angular momentum results in zero magnetic and electric moments. Consequently, odd-Z-even-N, or even-Z-odd-N nuclei to some degree, may be considered to consist of an even-Z-even-N core plus an unpaired nucleon. Most of the properties of these nuclides are then determined largely by the state of the unpaired nucleon. In odd-Z-odd-N nuclides the unpaired proton and neutron determine the characteristics of the nuclei.

Along with energy and angular momentum, **parity** is one of the essential parameters needed to specify the state of a nucleus. Parity has to do with the mathematical operation of projecting every point of a system through the origin of the coordinate axes to the opposite side. That is, x is transformed to −x, y to −y, and z to −z so that the wave function Ψ (x, y, z, t) becomes Ψ (−x, −y, −z, t). Now, one can show that when this operation does not change the form of the potential energy U, i.e., when

$$U(x, y, z) = U(-x, -y, -z) \tag{62}$$

then either of two things happens to the wave function:

1. Ψ will be totally unchanged, in which case its parity is said to be "even".
2. Ψ will change sign, in which case its parity is said to be "odd".

In either case, the probability density $|\Psi|^2$ remains the same, signifying that the essential physics of the situation has not changed as a result of the parity operation. Furthermore, if Equation 62 is true, the wave function of an isolated system having a definite parity (odd or even) will always retain the same parity; i.e., parity will be conserved. It happens that all forces in nature are parity-conserving with the exception of the weak interaction. Now, except in the process of beta decay, the weak interaction plays a minor role compared to the strong interaction and so the nucleus can be assumed to have states of definite parity.

With reference to our simple picture of the nucleus, parity is related to the ℓ quantum number of the nucleons in the nucleus. The reason is that for spherically symmetric potentials it can be shown that the parity of the wave function of a particle is always even when ℓ is even and is odd when ℓ is odd. Since the nucleons are considered to move independently of one another, we can write the total wave function Ψ_{total} of the nucleus as the product of the individual wave functions of each nucleon (or as a linear combination of these products). Then the parity of Ψ_{total} is given by the product of the parities of the individual wave functions and it is seen that the parity of the nucleus is even when the sum of the individual ℓ-values is even, and the parity is odd when the sum of the individual ℓ-values is odd. For an even-Z-even-N nucleus in its ground state, the sum of the ℓ-values is zero so that the nuclear parity is even. For a nucleus having an odd number of neutrons or protons, the parity of the nucleus is determined by the ℓ quantum number of the odd nucleon.

The tendency for stable nuclides to have about as many protons as neutrons is also explained in part by the shell model. Nuclear stability is associated with a nuclide having a lower energy than that of any neighbor to which it might decay. Since the energy levels for neutrons and protons are approximately the same, the lowest total energy for a nucleus of mass number A occurs when the lowest possible energy levels have been filled first, i.e., when $Z \approx N \approx A/2$. Of course, for high-Z nuclides Coulomb repulsion "pushes up" the energy levels of protons relative to neutrons so that the lowest energy nuclides in this case have more neutrons than protons.

Unfortunately, the shell model of the nucleus is not as successful in predicting nuclear properties as is the Hartree theory in predicting atomic properties. For example, the excited states of nuclei are only very approximately accounted for by the shell model. Neither are nuclear magnetic dipole moments or electric quadrupole moments accurately predicted. Experimental evidence indicates that, rather than moving independently, nucleons interact with one another in such a way as to move cooperatively to some extent. This effect, which leads to nonspherical nuclear shapes and to vibrational and rotational motions, is taken into consideration in collective models of the nucleus described in Volume II, Chapter 1.

Chapter 2

ATOMIC STRUCTURE

R. R. Sharma

TABLE OF CONTENTS

I. INTRODUCTION

The knowledge of the structure of atoms is of fundamental importance for understanding not only the structure of matter but also various physical processes in the universe. Electrons, X-rays, and α particles were discovered in the late 1890s and early 1900s and were used in scattering experiments to produce important and surprising data on the structure of matter. The experimental results obtained by Geiger and Marsden from the scattering of α particles by metal foils led Rutherford to postulate the "nuclear atom" in which the positive charge of the atom was assumed to be concentrated in a very small region, about 10^{-12} cm in radius at the center of the atom, with the negative charge diffused outside in a sphere of about 10^{-8} cm in radius. Before this deduction, the Thomson model of the atom was prevalent. According to this model, the positive charge of the atom was uniformly distributed throughout the atomic sphere. The fact that most of the mass of the atom is associated with the positive charge was also recognized from the experimental observation of the electromagnetic deflection of ions and electrons. The Rutherford model, in which the positive charge and the mass of the atom are centered in the small region called the nucleus, was able to explain the experimental scattering data and was widely accepted.

From optical spectroscopy the characteristic radiations of various atoms were also known. The arc or spark spectra of various atoms and ions gave rise to abundant "emission lines" with characteristic radiation frequencies. Balmer in 1885 established that the emission lines of hydrogen could be represented by a series, and within each series the wavelength λ of the emitted lines could be expressed by

$$\lambda = \frac{an^2}{(n^2 - b^2)}$$

with n as an integer and a and b as constants. Subsequently, Rydberg generalized the above formula in the form

$$\frac{\nu_m}{c} = \frac{\nu_\infty}{c} = \frac{R}{(m + \mu)^2}$$

where c is the velocity of light, ν_m is the frequency of various lines in a given series which converges to the frequency ν_∞, μ is an integer corresponding to a spectral series within the total spectrum of an atom, and m assumes the integral values representing successive members of the series. R is known as the Rydberg constant, which is always the same for a given element and varies only slightly from one element to another.

There remained the problem of explaining the optical spectra of atoms. The Thomson model, though it predicted a frequency of oscillation of an electron in the optical region, could not explain the series of spectral lines of the hydrogen atom. In the Rutherford model of the atom the question arose as to the location of the electron in the atom. Because of the electrostatic interaction between the central positive nucleus and the electron, according to classical mechanics the electron would fall into the nucleus, into a radius of about 10^{-12} cm. This led to the same situation as in the Thomson model, with the added difficulty as regards the size of the atom. The dilemma of explaining the atomic spectra still remained.

In 1912 Bohr proposed a model that resolved the difficulties with amazing success. The Bohr model assumes that an electron in an atom occupies specific planetary orbits which are stationary, that is, their energies remain constant. The electrons moving in these orbits do not radiate electromagnetic radiation though classically they would radiate energy continuously because of their spiraling motion around the positive nucleus. Bohr utilized Planck's hypothesis that matter radiates energy in the form of quanta of energy $h\nu$ (where h is Planck's constant and ν is the frequency of the electromagnetic radiation) and showed that the nonradiating states can be derived by quantization of the angular momentum of the electron in the form $n\, h/2\pi$, where n is an integer. According to the Bohr model the electron can jump from one orbit to the other giving rise to electromagnetic radiation of frequency

$$\nu = \frac{E_2 - E_1}{h}$$

where E_2 and E_1 are the energies of the two states. He deduced that the energy E_n of the stationary orbital states is given by

$$E_n = -\frac{2\pi^2 me^4 z^2}{h^2}\frac{1}{n^2}$$

where Ze is the nuclear charge, m is the electronic mass, and e the magnitude of the electronic charge. For the energy difference between two levels, designated by n_1 and n_2, the above expression yields

$$E_{n_2} - E_{n_1} = K\left(\frac{1}{n_2^2} - \frac{1}{n_1^2}\right)$$

where

$$K = \frac{2\pi^2 me^4 Z^2}{h^2}$$

These expressions immediately produce the Balmer formula or the Rydberg formula. Thus the Bohr model was able to explain qualitatively as well as quantitatively the characteristic spectra of atoms, though there still remained some unexplained features such as the multiplicity of spectral lines. Next, Sommerfeld improved the Bohr model by making use of the generalized Bohr quantization rule (known as the Wilson-Sommerfeld quantization rule) and introduced elliptical stationary orbits to explain the fine structure of the hydrogen spectrum.

In 1925 Schrödinger and simultaneously Born and Heisenberg proposed rigorous quantum mechanical theories which, though seemingly different, turned out to be equivalent. Schrödinger identified the similarities between the eigenvalues and eigenfunctions of differential equations and the "stationary states" of quantum mechanics. He formulated a self-consistent theory known as wave mechanics and the famous equation which now bears his name. Born and Heisenberg recognized the similarities between the nature of the dynamical variables and of certain linear operators (and associated matrices) to formulate their matrix mechanics. The Schrödinger treatment is simpler and conceptually easier though the matrix formulation is certainly elegant and powerful. In the present review, we shall utilize only the Schrödinger treatment.

There are now excellent books[1-22] available which deal with the quantum mechanical theory at the lower as well as higher levels. In this article we shall systematically present the quantum mechanical developments for atoms containing one or more than one electron and consider the details of the spectra of atoms. In Section II we treat one-electron atoms and in Section III multielectron atoms. In these sections we also analyze various important interactions and processes, supplying appropriate theoretical foundations. Section III also reviews various theoretical methods that are useful for calculating the electronic structure of multielectron atoms. Characteristic X-rays and optical excitations are considered in Section IV and the Auger Effect in Section V. In Section VI we present the concluding remarks.

II. QUANTUM MECHANICS OF ONE-ELECTRON ATOMS

A. Introduction

The Schrödinger theory of the one-electron atom is of utmost importance since it lays down the foundations of the quantum mechanical treatment of multielectron atoms, molecules, nuclei, and solids. The simplest one-electron system is the hydrogen atom, which is also of great historical value owing to the fact that Schrödinger treated this system first by his quantum mechanical theory. The Schrödinger theory not only predicts the energy eigenvalues but also gives the wave functions which, in turn, are helpful in obtaining various properties of atoms such as the probability density functions, the orbital angular momenta, the expectation values of spin-dependent effects, the transition rates for optical absorptions, and so on.

The one-electron atom constitutes a three-dimensional case of a two-particle system consisting of a positively charged nucleus and a negatively charged electron bound together by the attractive electrostatic coulomb force (due to the charges on the particles). This is one of the few systems that can be solved exactly. Because it serves as one of the examples of a larger class of systems involving a central force, it is an important problem from a pedagogical point of view. By a central force we mean that it is a force which is directed radially inward (or outward) from the origin and varies in magnitude as a function of radial distance measured from the origin.

As will be clear in the following discussion, the one-electron atom containing two particles causes no difficulty if one makes use of the reduced mass technique. This technique effectively transforms this problem into an equivalent system of a nucleus with infinite mass and an electron with the reduced mass μ given by

$$\mu = \frac{M}{m + M} m \tag{1}$$

where m and M are the true masses of the electron and nucleus, respectively. The electron with mass μ moves around the infinitely massive nucleus retaining the same electron-nucleus distance as in the actual atom. Thus the transformed problem is a simplified one in which only a single particle moves instead of a pair of particles being in motion.

For the solution of the Schrödinger equation of a one-electron atom it is useful to adopt the method of separation of variables, according to which the solution is assumed to be a product of functions, each being dependent only on a single, independent variable. Since the Coulomb potential between the charged particles depends on the distance between the particles, it is convenient to make use of the spherical coordinates as the three independent variables and each of the three product functions as dependent only on one of the three spherical coordinates of the electron with respect to the nucleus. It is then straightforward to separate the Schrödinger equation into three simpler equations, each dependent on only one of the independent variables. Next, these equations are solved individually to yield information as to the complete solution of the Schrödinger equation.

B. Schrödinger Equation

We start with the actual problem consisting of a nucleus of mass M and an electron mass m in a given frame of reference (with origin O′) as shown in Figure 1. The Hamiltonian operator $\mathcal{H}'$ of the system is the sum of the kinetic energy operators of the nucleus of charge Ze and the electron (with charge −e), plus the potential energy of interaction. Here we have assumed that, in general, the nuclear charge Ze is different than the number of electrons so that the derived results could very well be applied, not only to the neutral hydrogen atom (Z = 1), but also to single-electron ionized atoms; for instance for He^{+} (Z = 2), Li^{2+} (Z = 3), and so on. Thus

$$\begin{aligned} \mathcal{H}' = &-\frac{\hbar^2}{2M}\left\{\frac{\partial^2}{\partial X'^2} + \frac{\partial^2}{\partial Y'^2} + \frac{\partial^2}{\partial Z'^2}\right\} \\ &-\frac{\hbar^2}{2m}\left\{\frac{\partial^2}{\partial x'^2} + \frac{\partial^2}{\partial y'^2} + \frac{\partial^2}{\partial z'^2}\right\} \\ &+ V(X', Y', Z';\ x', y', z') \end{aligned} \tag{2}$$

where X′, Y′, Z′, and x′, y′, z′, are the coordinates of the nucleus and the electron, respectively, and

$$V(X', Y', Z',;\ x', y', z') = -\frac{1}{4\pi\epsilon_0}\frac{Ze^2}{\sqrt{(X'-x')^2 + (Y'-y')^2 + (Z'-z')^2}} \tag{3}$$

The time-dependent Schrödinger equation of the system is then

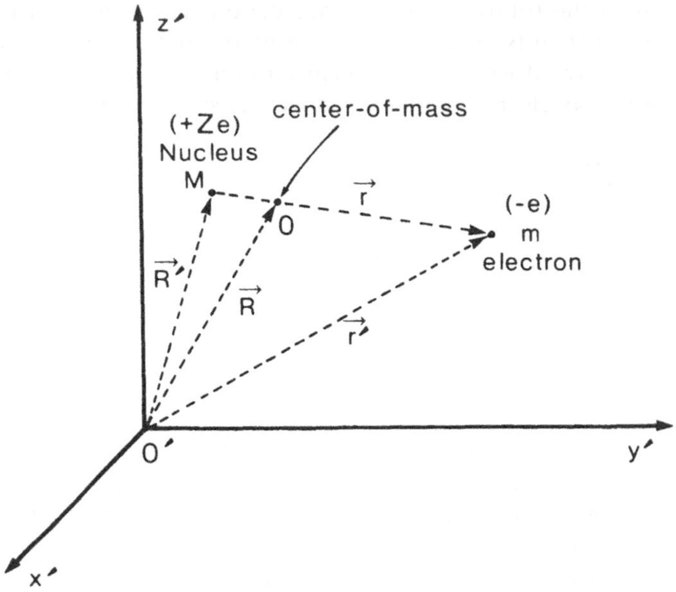

FIGURE 1. Depiction of nucleus of mass M and charge Ze and electron of mass m and charge (−e) forming a single-electron atom. The center of mass of the system is denoted by O. The origin of the coordinate system is at O′. The coordinates of the nucleus are X′, Y′, Z′ and those of the electron, x′, y′, z′; $\vec{R}' = (X', Y', Z')$ and $\vec{r}' = (x', y', z')$.

$$\mathcal{H}'\Psi'(X', Y', Z';\ x', y', z';\ t) = i\hbar \frac{\partial}{\partial t} \Psi'(X', Y', Z';\ x', y', z';\ t) \tag{4}$$

where $\mathcal{H}'$ is given by Equations 2 and 3 and Ψ' (X′, Y′ Z′; x′, y′, z′t) is the wave function. Since the potential V (X′, Y′, Z′; x′, y′, z′) is independent of time t, it is possible to use the technique of the separation of variables to separate Equation 4 into two equations, one depending on the space variables X′, Y′, Z′ and x′, y′, z′, and the other on t. For this purpose, one uses

$$\Psi'(X', Y', Z';\ x', y', z';\ t) = \Psi(X', Y', Z';\ x', y', z')\ \Phi(t) \tag{5a}$$

or, in short notation

$$\Psi'(\vec{R}';\vec{r}', t) = \Psi(\vec{R}';\vec{r}')\ \Phi(t) \tag{5b}$$

Making use of Equation 5 in Equation 4, one then obtains the time-independent Schrödinger equation,

$$\left\{ -\frac{\hbar^2}{2M} \nabla'^2_M - \frac{\hbar^2}{2m} \nabla'^2_m + V(|\vec{R}' - \vec{r}'|) \right\} \Psi(\vec{R}', \vec{r}') = E'\ \Psi(\vec{R}', \vec{r}') \tag{6}$$

and the time-dependent equation,

$$i\hbar \frac{\partial \Phi(t)}{\partial t} = E' \frac{\partial \Phi(t)}{\partial t} \quad (7)$$

where

$$\nabla_M'^2 = \frac{\partial^2}{\partial X'^2} + \frac{\partial^2}{\partial Y'^2} + \frac{\partial^2}{\partial Z'^2} \quad (8)$$

$$\nabla_m'^2 = \frac{\partial^2}{\partial x'^2} + \frac{\partial^2}{\partial y'^2} + \frac{\partial^2}{\partial z'^2} \quad (9)$$

and

$$V = -Ze^2/4\pi\epsilon_0 \, |\vec{R}' - \vec{r}'| \quad (10)$$

E' has entered into the equations as a constant of the separation of variables. The solution of Equation 7 is

$$\Phi(t) = A' e^{\frac{i}{\hbar} E' t} \quad (11)$$

where A' is a constant of integration. It is because of the solution (Equation 11) and the Planck-Einstein relation,

$$\nu = \frac{E'}{h} \quad (12)$$

that the constant E' can be identified as the energy of the system.

Equation 6 contains six space variables X', Y', Z', and x', y', z'. Since the potential V (see Equation 10) is a function of $|\vec{R}' - \vec{r}'|$, Equation 6, as it is, cannot be separated into two or more separate equations. One therefore transforms the coordinates into the set of new space variables, which corresponds to taking the origin of the coordinate system at the center of mass of the system and measuring the coordinates of the electron relative to the nucleus. Thus, denoting the new coordinates as X, Y, Z, and x, y, z, we have,

$$X = \frac{MX' + mx}{M+m} \; ; \quad x = X' - x'$$

$$Y = \frac{MY' + my}{M+m} \; ; \quad y = Y' - y'$$

$$Z = \frac{MZ' + mz}{M+m} \; ; \quad z = Z' - z' \quad (13)$$

Making use of these transformations, one gets from Equation 6

$$\left\{ \frac{-\hbar^2}{2(M+m)} \left(\frac{\partial^2}{\partial X^2} + \frac{\partial^2}{\partial Y^2} + \frac{\partial^2}{\partial Z^2} \right) - \frac{\hbar^2}{2\mu} \left(\frac{\partial^2}{\partial x^2} + \frac{\partial^2}{\partial y^2} + \frac{\partial^2}{\partial z^2} \right) - \frac{Ze^2}{4\pi\epsilon_o r} \right\} \Psi(\vec{R}, \vec{r})$$

$$= E'\Psi(\vec{R}, \vec{r}) \tag{14}$$

where $\vec{R} = (X, Y, Z)$ and $\vec{r} = (x, y, z)$. $\Psi(\vec{R}, \vec{r})$ can be obtained from $\Psi(\vec{R'}, \vec{r'})$ by using the transformation of the coordinates given in Equation 13. One again makes use of the separation of variables technique by writing

$$\Psi(\vec{R}, \vec{r}) = \psi(\vec{r})\, \chi(\vec{R}) \tag{15}$$

to obtain from Equation 14

$$\left[-\frac{\hbar^2}{2\mu} \nabla^2 - \frac{Ze^{2\prime}}{4\pi\epsilon_o r} \right] \psi(\vec{r}) = E\psi(\vec{r}) \tag{16}$$

and

$$\frac{-\hbar^2}{2(M+m)} \nabla_T^2\, \chi(\vec{R}) = (E' - E)\, \chi(\vec{R}) \tag{17}$$

where

$$\nabla^2 = \frac{\partial^2}{\partial x^2} + \frac{\partial^2}{\partial y^2} + \frac{\partial^2}{\partial z^2} \tag{18}$$

and

$$\nabla_T^2 = \frac{\partial^2}{\partial X^2} + \frac{\partial^2}{\partial Y^2} + \frac{\partial^2}{\partial Z^2} \tag{19}$$

Equation 16 describes the motion of the electron relative to the nucleus since it involves only the relative coordinates x, y, z. On the other hand, Equation 17 describes the motion of the center of mass. The solution of Equation 17 is

$$\chi(\vec{R}) = N\, e^{\pm i \vec{K} \cdot \vec{R}} \tag{20}$$

which is usually referred to as the free-particle wave function (corresponding to the particle in zero potential) where $\vec{K}$ is the wave vector with magnitude,

$$K = \frac{2}{\sqrt{\hbar^2}} (E' - E)\ (M+m)$$

This gives the total energy of the system as

$$E' = E + \frac{\hbar^2 K^2}{2(M+m)} \tag{21}$$

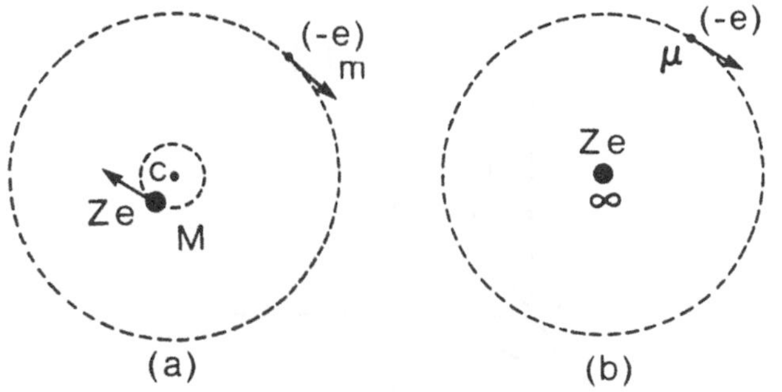

FIGURE 2. (a) One-electron atom in which the nucleus of mass M and charge Ze and the electron of mass m and charge (−e) have been shown to move about their fixed center of mass C. (b) One-electron atom in an equivalent model representation in which a particle of charge (−e) and reduced mass μ moves around a fixed nucleus of charge (Ze).

which is the sum of the kinetic energy of the center of mass and the energy of the electron and the nucleus in the center of mass reference frame. If the center of mass is at rest then $E' = E$, and Equation 16 then constitutes the Schrödinger equation of the system. The same equation is obtained if one writes the Schrödinger equation for a system of an electron of mass μ and charge −e, and a nucleus of infinite mass and charge Ze. It is because of this fact that one usually writes the Schrödinger equation directly in the form of Equation 16, using the reduced mass μ given by Equation 1 (see also Figure 2). One notes that since the nuclear mass M is about 2000 times larger than the electron mass, $\mu \simeq m$. In that case, the time-independent Schrödinger equation reads

$$\left(\frac{-\hbar^2}{2m} \nabla^2 - \frac{Ze^2}{4\pi\epsilon_0 r} \right) \psi(\vec{r}) = E\psi(\vec{r}) \tag{22}$$

which describes an electron of mass m and charge −e interacting with a nucleus of infinite mass and charge Ze. $\psi(\vec{r})$ is the eigenfunction of the electron.

Equation 16 is a three-dimensional equation. Since the potential in the present case is spherically symmetric and central, it is possible to separate Equation 16 into three independent equations corresponding to the three electron coordinates r, θ, and ϕ (see Figure 3). One notes that this separation is not possible if one treats the problem in the Cartesian reference frame since the potential is then in a complicated form (namely, $1/\sqrt{x^2 + y^2 + z^2}$). Accordingly, using the product form of the wave function, we have

$$\psi(\vec{r}) = R(r)\ Y(\theta, \phi) \tag{23}$$

where R(r) is a function of r alone and $Y(\theta,\phi)$ is a function of θ and ϕ only. Making use of the form of ∇^2 in the spherical coordinates, viz.,

$$\nabla^2 = \frac{1}{r^2}\frac{\partial}{\partial r} r^2 \frac{\partial}{\partial r} + \frac{1}{r^2 \sin\theta}\frac{\partial}{\partial\theta} \sin\theta \frac{\partial}{\partial\theta} + \frac{1}{r^2 \sin^2\theta}\frac{\partial^2}{\partial\phi^2} \tag{24}$$

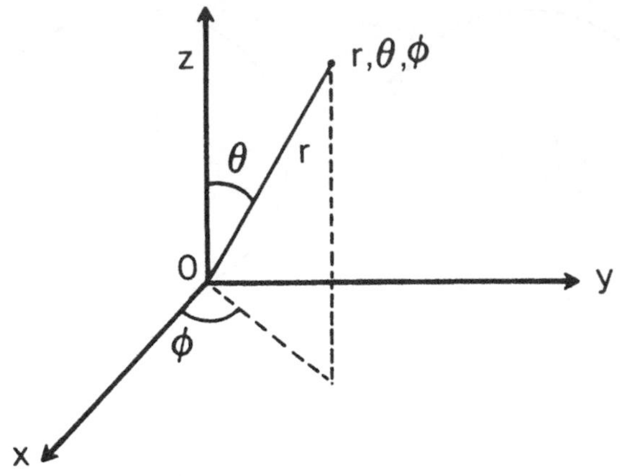

FIGURE 3. Depiction of the electron coordinates r, θ, φ, with the nucleus located at the origin O.

in Equation 16 together with the product form given in Equation 23, one separates Equation 16 into the following two equations:

$$\left[\frac{1}{r^2}\frac{\partial}{\partial r}\,r^2\,\frac{\partial R}{\partial r} + \frac{2M}{\hbar^2}\,(E-V(r))\right] R(r) = \frac{\lambda}{r^2}\,R(r) \tag{25}$$

and

$$\left[\frac{1}{\sin\theta}\,\frac{\partial}{\partial\theta}\,\sin\theta\,\frac{\partial}{\partial\theta} + \frac{1}{\sin^2\theta}\,\frac{\partial^2}{\partial\phi^2} + \lambda\right] Y(\theta,\phi) = 0 \tag{26}$$

where

$$V(r) = -\frac{Ze^2}{4\pi\epsilon_o r} \tag{27}$$

and λ is the separation constant. Similarly, one assumes

$$Y(\theta,\phi) = \Theta(\theta)\,\Phi(\phi) \tag{28}$$

to separate Equation 26 into two equations, one depending on θ alone,

$$\left(\frac{1}{\sin\theta}\,\frac{\partial}{\partial\theta}\,\sin\theta\,\frac{\partial}{\partial\theta} + \lambda - \frac{\beta}{\sin^2\theta}\right)\Theta(\theta) = 0 \tag{29}$$

and the other on ϕ alone,

$$\frac{d^2\Phi(\phi)}{d\phi^2} = -\beta\Phi(\phi) \qquad (30)$$

with β as the separation constant.

From the nature of Equations 25 to 30 it is obvious that for any central force problem, the functions $\Theta(\theta)$ and $\Phi(\phi)$ remain the same. The radial function R(r), however, depends on the given problem because of the characteristic potential V(r). Associated with the problem, λ and β in Equations 25 and 30 are related to the constants of motion, viz., the angular momentum and its projection along a given axis, as will be clear shortly.

C. Solution of the θ and ϕ Parts of the Schrödinger Equation

The ϕ equation is given by Equation 30 which has the particular solution

$$\Phi(\phi) = A\, e^{i\sqrt{\beta}\,\phi} \qquad (31)$$

where A is a constant. The solution of the form given in Equation 31 has been taken so as to maintain the probability density $\Phi^*\Phi$ independent of the angle ϕ, which must be the case since the potential V(r) is spherically symmetric. The potential depends only on the distance from the electron to the nucleus and consequently none of the properties should single out any particular direction in space because all directions are equivalent. In fact, if the potential has an axial rotational symmetry, the functions of Equation 31 can still be used. However, if no axial symmetry is present, as in the case of molecules, one frequently uses solutions of the form $\cos\sqrt{\beta}\phi$ and $\sin\sqrt{\beta}\phi$.

Considering the explicit requirement that the wave function be single-valued, we impose the condition on Equation 31 that

$$\Phi(\phi) = \Phi(2\pi + \phi) \qquad (32)$$

which yields

$$e^{i\sqrt{\beta}\,2\pi} = 1 \qquad (33)$$

or

$$\sqrt{\beta} = \pm m \qquad (34)$$

where m is an integer. Using the quantum number m as the subscript, we rewrite Equation 31 as

$$\Phi_m(\beta) = A_m\, e^{im\phi} \qquad (35)$$

where now m assumes negative as well as positive integer values.

The constant A_m is determined from the normalization requirement,

$$\int_0^{2\pi} \Phi_m^*(\phi)\Phi_m(\phi)\, d\phi = 1 \qquad (36)$$

which yields

$$A_m = \frac{1}{\sqrt{2\pi}} \tag{37}$$

neglecting a constant phase factor. The quantum number m is known as the magnetic quantum number, whose significance will be clear when we consider the Zeeman effect (see Section IV.F).

The solution of the θ equation (Equation 29) cannot be obtained quite so easily. One first uses the new variable Z where

$$Z = \cos\theta \tag{38}$$

so that

$$P(Z) = \Theta(\theta) \tag{39}$$

to transform Equation 29 into the form

$$\left[\frac{\partial}{\partial Z}(1 - Z^2)\frac{\partial}{\partial Z} + \left(\lambda - \frac{m^2}{1-Z^2}\right)\right] P(Z) = 0 \tag{40}$$

The domain of Z is

$$-1 \leqslant Z \leqslant 1 \tag{41}$$

since θ varies from 0 to π. We seek solutions which are single-valued, finite, and continuous throughout the domain of Z. For m = 0, Equation 40 reduces to the simpler equation,

$$\left[\frac{\partial}{\partial Z}(1 - Z^2)\frac{\partial}{\partial Z} + \lambda\right] P(Z) = 0 \tag{42}$$

which is known as Legendre's equation. To solve it one assumes that

$$P(Z) = \Sigma_{n'}\, a_{n'} Z^{n'} \tag{43}$$

Substituting the value of P(Z) from Equation 43 in Equation 42, one deduces the recursion relation (on collecting the coefficients of Z^{ℓ} from both sides),

$$a_{\ell+2} = \frac{\ell(\ell+1) - \lambda}{(\ell+2)(\ell+1)}\, a_\ell \tag{44}$$

Obviously,

$$\text{Limit}_{\ell\rightarrow\infty} \left\{\frac{a_{\ell+2}}{a_\ell}\right\} = 1 \tag{45}$$

Therefore the series (Equation 43) diverges for Z = 1 unless it is terminated. The term containing $Z^{\ell+2}$ vanishes if

$$\lambda = \ell(\ell+1) \tag{46}$$

In that case the polynomial P(Z) is of order ℓ and we denote it by $P_\ell(Z)$ with the subscript ℓ to explicitly reveal its order. There exist other solutions known as $Q_\ell(Z)$ of the θ equation (Equation 29) which are really the polynomials of order ℓ in 1/Z. Since $Q_\ell(Z)$ are not finite throughout the domain of Z, they are not physically acceptable solutions of the Schrödinger equation.

Equation 44 provides the values of the coefficients appearing in Equation 43 which, in turn, leads to the expressions for $P_\ell(Z)$ within a multiplicative constant. It can be shown (by direct substitution) that

$$P_\ell^m(Z) = (1 - Z^2)^{m/2} \frac{d^m}{dZ^m} P_\ell(Z) \tag{47}$$

is the solution of Equation 40. Equation 40 is known as the associated Legendre equation and $P_\ell^m(Z)$ the associated Legendre functions. Equation 47 restricts the m values,

$$m \leqslant \ell \tag{48}$$

Thus, the constants m, λ, and ℓ have the values

$$m = 0, 1, 2, \ldots \ell$$

$$\lambda = \ell(\ell+1)$$

$$\ell \geqslant m$$

Some of the values of the Legendre functions P_ℓ^m have been listed in Table 1.

The product of the functions $P_\ell^m(\cos\theta)$ and $\Phi_m(\phi)$ are the well-known spherical harmonics $Y_\ell^m(\theta, \phi)$ given by

$$Y_\ell^{\pm m} = N_\ell^{\pm m} P_\ell^m(\cos\theta)\, e^{\pm im\phi} \tag{49}$$

$N^{\pm m}$ are the normalization constants obtained by the normalization condition

$$\int_{\phi=0}^{2\pi} \int_{\theta=0}^{\pi} Y_\ell^{m*}(\theta, \phi)\, Y_\ell^m(\theta,\phi) \sin\theta \, d\theta \, d\phi = 1 \tag{50}$$

which gives

$$N_\ell^{\pm m} = \left(\frac{(2\ell + 1)}{4\pi} \frac{(\ell - m)!}{(\ell + m)!} \right)^{1/2} \tag{51}$$

The spherical harmonics then satisfy the following general orthonormalization conditions,

$$\int Y_\ell^{m*}(\theta, \phi)\, Y_\ell^m(\theta,\phi) \sin\theta \, d\theta \, d\phi = \delta_{\ell,\ell'} \delta_{mm'} \tag{52}$$

where, in general, $\delta_{p,p}$ is the Krönecker delta function defined by

$$\delta_{p,p'} = \begin{cases} 1 \text{ if } p = p' \\ 0 \text{ if } p \neq p' \end{cases}$$

Table 1
LIST OF THE LEGENDRE FUNCTIONS $P_\ell^m(Z)$ FOR A SET OF VALUES OF ℓ AND m

	$P_\ell^m(Z)$			
ℓ/m	0	1	2	3
0	1			
1	Z	$\sqrt{1-Z^2}$		
2	$\frac{1}{2}(3Z^2-1)$	$3Z(1-Z^2)^{1/2}$	$3(1-Z^2)$	
3	$\frac{1}{2}(5Z^3-3Z)$	$3/2(5Z^2-1)(1-Z^2)^{1/2}$	$15Z(1-Z^2)$	$15(1-Z^2)^{3/2}$

Table 2
LIST OF Y_ℓ^m FUNCTIONS

ℓ	m	$Y_\ell^m(\theta,\phi)$
0	0	$Y_0^0 = \frac{1}{\sqrt{4\pi}}$
1	0	$Y_1^0 = \sqrt{\frac{3}{4\pi}}\cos\theta$
1	±1	$Y_1^{\pm1} = \mp\frac{\sqrt{3}}{\sqrt{8\pi}}\sin\theta\, e^{\pm i\phi}$
2	0	$Y_2^0 = \sqrt{5/16\pi}\,(3\cos^2\theta-1)$
2	±1	$Y_2^{\pm1} = \mp\left(\frac{15}{8\pi}\right)^{1/2}\sin\theta\cos\theta\, e^{\pm i\phi}$
2	±2	$Y_2^{\pm2} = (15/32\pi)^{1/2}\sin^2\theta\, e^{\pm 2i\phi}$
3	0	$Y_3^0 = (7/16\pi)^{1/2}(5\cos^3\theta-3\cos\theta)$
3	±1	$Y_3^{\pm1} = \mp(21/64\pi)^{1/2}\sin\theta(5\cos^2\theta-1)\, e^{\pm i\phi}$
3	±2	$Y_3^{\pm2} = (105/32\pi)^{1/2}\sin^2\theta\cos\theta\, e^{\pm 2i\phi}$
3	±3	$Y_3^{\pm3} = \mp(35/64\pi)^{1/2}\sin^3\theta\, e^{\pm 3i\phi}$

In Table 2 the values of the normalized Y_ℓ^m for $\ell \leqslant 3$ have been compiled.

D. The Angular Momentum and Angular Momentum Eigenfunctions

For a particle with the position vector $\vec{r}$ and linear momentum $\vec{p}$, the angular momentum $\vec{L}$ is defined by

$$\vec{L} = \vec{r} \times \vec{p} \qquad (53)$$

It can be shown that the angular momentum of a closed system in an isotropic space is conserved (does not change). Substituting the quantum mechanical expression for the momentum operator $\vec{p}_{op}$, that is,

$$\vec{p}_{op} = \frac{\hbar}{i} \nabla \tag{54}$$

in Equation 53 one obtains the expressions for the components of $\vec{L}_{op}$, the angular momentum operator,

$$L_{x_{op}} = \frac{\hbar}{i}\left(y\frac{\partial}{\partial z} - z\frac{\partial}{\partial y}\right)$$

$$L_{y_{op}} = \frac{\hbar}{i}\left(z\frac{\partial}{\partial x} - x\frac{\partial}{\partial z}\right)$$

$$L_{z_{op}} = \frac{\hbar}{i}\left(x\frac{\partial}{\partial y} - y\frac{\partial}{\partial x}\right) \tag{55}$$

If the system is subjected to an external field the angular momentum is, in general, not conserved. However, if the system is in a centrally symmetric field, the angular momentum is still a conserved quantity with respect to the center. Also, for an axial symmetric potential the angular moment about the symmetry axis is conserved. One notes that these laws are valid in classical mechanics as well as in quantum mechanics.

If the angular momentum is conserved for a system, the system has a definite value of the angular momentum in the stationary state. Consequently, for a stationary state it is of importance to know the average value of the angular momentum. It should be remarked that the average value of the angular momentum is zero for a nondegenerate state, that is, for a state that has only one eigenfunction for a given energy.

Defining, in general, the commutator between the operators A and B as

$$[A,B] = AB - BA \tag{56}$$

one can verify that by using Equation 55

$$[L_{x_{op}}, L_{y_{op}}] = i\hbar L_{z_{op}}$$

$$[L_{y_{op}}, L_{z_{op}}] = i\hbar L_{x_{op}}$$

$$[L_{z_{op}}, L_{x_{op}}] = i\hbar L_{y_{op}} \tag{57}$$

Further, we define the square of the angular momentum,

$$L_{op}^2 = L_{x_{op}}^2 + L_{y_{op}}^2 + L_{z_{op}}^2 \tag{58}$$

which satisfies the commutation rules

$$[L_{op}^2, L_{x_{op}}] = 0 \tag{59}$$

$$[L_{op}^2, L_{y_{op}}] = 0 \tag{60}$$

$$[L_{op}^2, L_{z_{op}}] = 0 \tag{61}$$

The Cartesian coordinates are related to the spherical coordinates (see Figure 3) by

$$x = r \sin\theta \cos\phi$$

$$y = r \sin\theta \sin\phi$$

$$z = r \cos\theta \tag{62}$$

One can prove, with the help of Equation 62, that

$$\frac{\partial}{\partial\phi} = x \frac{\partial}{\partial y} - y \frac{\partial}{\partial x} \tag{63}$$

Combining Equations 55 and 63, one then gets the operator form of the z-component of the angular momentum in the spherical coordinate system,

$$L_{z_{op}} = \frac{\hbar}{i} \frac{\partial}{\partial\phi} \tag{64}$$

Similarly, using Equations 55, 58, and 62 we have

$$L_{x_{op}} = \frac{\hbar}{i} \left(\sin\phi \frac{\partial}{\partial\theta} + \cot\theta \cos\phi \frac{\partial}{\partial\phi} \right) \tag{65}$$

$$L_{y_{op}} = \frac{\hbar}{i} \left(\cos\phi \frac{\partial}{\partial\theta} - \cot\theta \sin\phi \frac{\partial}{\partial\phi} \right) \tag{66}$$

$$L^2_{op} = -\hbar^2 \left[\frac{1}{\sin\theta} \frac{\partial}{\partial\theta} \sin\theta \frac{\partial}{\partial\theta} + \frac{1}{\sin^2\theta} \frac{\partial^2}{\partial\phi^2} \right] \tag{67}$$

From Equations 35 and 64,

$$L_{z_{op}} \Phi_m = m \hbar \Phi_m \tag{68}$$

and

$$L^2_{z_{op}} \Phi_m = m^2 \hbar^2 \Phi_m \tag{69}$$

which show that the eigenvalue of the z-component of the angular momentum is $m\hbar$. Since m is an integer, Equation 68 shows that the z-component of the angular momentum is quantized.

In view of Equations 26, 28, 29, 30, 35, 46, 49, and 67, one deduces

$$L^2_{op} Y^m_\ell(\theta,\phi) = -\hbar^2 \left(\frac{1}{\sin\theta} \frac{\partial}{\partial\theta} \sin\theta \frac{\partial}{\partial\theta} + \frac{1}{\sin^2\theta} \frac{\partial^2}{\partial\phi^2} \right) Y^m_\ell(\theta,\phi)$$

$$= \ell(\ell+1)\hbar^2 Y^m_\ell(\theta,\phi) \tag{70}$$

which gives the celebrated result that the eigenvalue of the square of the angular momentum is $\ell(\ell+1)\hbar^2$. ℓ is called the angular momentum quantum number or azimuthal quantum number. It is important to observe (from Equation 70) that the magnitude of the total angular momentum is not $\ell\hbar$ as Bohr had postulated, but $\sqrt{\ell(\ell+1)}\hbar$.

The restriction (see Equation 48)

$$m \leqslant \ell$$

is physically equivalent to the fact that the z-component of the angular momentum cannot exceed the magnitude of the total angular momentum. In fact, the expectation value of the z-component of the angular momentum cannot (except for $\ell = 0$ and m = 0) quite become equal to the magnitude of the angular momentum since

$$(L_z) = m\hbar \tag{71}$$

$$(L_z)_{max} = \ell\hbar \tag{72}$$

and

$$|L| = \sqrt{\ell(\ell+1)}\,\hbar \tag{73}$$

One notes from Equation 72 that the maximum allowed value of the component of L along the z-direction is $\ell\hbar$, a situation which is possible only when $\vec{L}$ is oriented in the z-direction. In that case the magnitude of $\vec{L}$, viz., $|L|$ is expected to be the same as $(L_z)_{max}$ and the values of the x- and y-components of $\vec{L}$ to be zero. However, according to Equation 73 $|L|$ is $\sqrt{\ell(\ell+1)}\,\hbar$ which is evidently not the same as $(L_z)_{max}$ (or $\ell\hbar$). In the wake of this it could be concluded that the values of the L_x and L_y components of $\vec{L}$ are not exactly zero. Thus, if the z-component of the angular momentum is known exactly, the values of the L_x and L_y components cannot be determined accurately at the same time. In fact, the minimum uncertainty product for the values of L_x and L_y can be shown to be

$$(\Delta L_x)\ (\Delta L_y) \geqslant \frac{\hbar}{2}$$

with similar uncertainty relations for other components of $\vec{L}$. These uncertainty relations are a direct consequence of the noncommutation relations given by Equation 57 between $L_{x\ op}$, $L_{y\ op}$ and $L_{z\ op}$. In general, if two operators P and Q do not commute (i.e., $[P,Q] \neq 0$), the corresponding observed quantities p and q associated with P and Q satisfy the uncertainty result

$$(\Delta p)\ (\Delta q) \geqslant \tfrac{1}{2}\,|\langle [P, Q] \rangle|$$

where the symbol $\langle A \rangle$, in general, represents the expectation value of A and the symbol $[A,B]$, in general, is the commutator of A and B as defined in Equation 56. Thus, Equations 71 to 73 constitute the direct manifestation of the minimum uncertainty product (generally known as the Heisenberg uncertainty product), a characteristic of noncommutating observables.

Though the z-component of the angular momentum is quantized, the x- and y-components are not. This is mysterious in view of classical mechanics. Classically, the angular momentum of a particle moving under the influence of a spherically symmetrical potential would be completely fixed in magnitude and direction and all three of its components would have definite values. In quantum mechanics the results are different because of the uncertainty principle, according to which no two components of an angular momentum can be known simultaneously with complete precision. Since the z-component of the angular momentum is known precisely to have the value $m\hbar$,

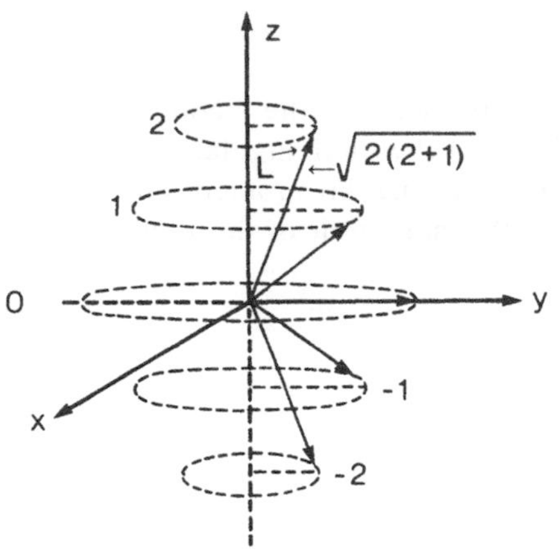

FIGURE 4. Representation of the angular momentum in a vector model for the special case of the quantum number $\ell = 2$. The magnitude of the angular momentum vector is $\sqrt{\ell(\ell+1)}$ or $\sqrt{2(2+1)}$ and the Z-components are +2, +1, 0, −1, −2 (all in units of $\hbar$).

the x- and y-components are indefinite. If one evaluates the expectation values of the x- and y-components of $\vec{L}$ one obtains zero results, which means that L_x and L_y fluctuate about their average value of zero. This represents the quantum mechanical picture of the conservation of the angular momentum.

The properties of the angular momentum can be represented by a vector model in which the magnitude of the angular momentum vector corresponding to the azimuthal quantum number ℓ is $\sqrt{\ell(\ell+1)}\,\hbar$ and the z-component of the angular momentum is $m\hbar$ with $m = \ell, (\ell-1), \ldots 0, \ldots -(\ell-1), (-\ell)$. The angular momentum vector precesses randomly about the z-axis, maintaining constant magnitude and a constant z-component. The vector model for the case of $\ell = 2$ has been illustrated in Figure 4. In the classical limit, $\ell \rightarrow \infty$ which gives $L = \sqrt{\ell(\ell+1)}\hbar \rightarrow \ell\hbar$. Thus in the state $m = \ell$, the vector $\vec{L}$ lies almost along the z-axis in the classical limit because the z-component is equal to $\ell\hbar$; in this case then $L_x = 0$ and $L_y = 0$. This conclusion agrees with the classical theory where all three components of the angular momentum can be determined simultaneously with complete precision.

The magnetic quantum number m shows the orientation in space of the angular momentum. In the case of an atom with an electron in a spherically symmetrical potential it also describes the orientation of the atom in space. If no external forces are present on the atom, the energy of the atom is not expected to depend on the orientation of the atom or, in other words, on the quantum number m. This indeed is the case as will be clear very soon when we shall obtain an expression for the energy. This also explains why the eigenfunctions are expected to be degenerate with respect to the quantum number m.

E. Solution of the Radial Equation

Substituting the value of λ from Equation 46 in Equation 25 we rewrite the radial Schrödinger equation in the form

$$\frac{1}{r^2}\frac{d}{dr}r^2\frac{dR(r)}{dr} - \frac{\ell(\ell+1)}{r^2}R(r) + \frac{2\mu}{\hbar^2}(E - V(r))\,R(r) = 0 \qquad (74)$$

We define now the new parameters α and n by

$$\alpha^2 = \frac{-8\mu E}{\hbar^2} \tag{75}$$

and

$$n = \frac{2\mu Ze^2}{4\pi\epsilon_0 \alpha \hbar^2} \tag{76}$$

In terms of these parameters, Equation 74 reads as

$$\frac{1}{r^2}\frac{d}{dr}r^2\frac{dR(r)}{dr} + \left(\frac{-\alpha^2}{4} + \frac{\alpha n}{r} - \frac{\ell(\ell+1)}{r^2}\right)R(r) = 0 \tag{77}$$

Converting to a dimensionless parameter ϱ where

$$\rho = \alpha r \tag{78}$$

the radial equation becomes

$$\frac{1}{\rho^2}\frac{d}{d\rho}\rho^2\frac{d}{d\rho}S(\rho) + \left(\frac{-1}{4} + \frac{n}{\rho} - \frac{\ell(\ell+1)}{\rho^2}\right)S(\rho) = 0 \tag{79}$$

with

$$S(\rho) = R(r) \tag{80}$$

The quantum mechanical requirements for the radial wave function $S(\varrho)$ or R(r) are that it should vanish sufficiently rapidly as r tends to infinity, should converge rapidly, and should be square integrable. It turns out that the solution of Equation 79 with the above properties is of the form

$$S(\rho) = e^{-\rho/2}\rho^{\ell}\sum_{\nu=0}^{n'}\alpha_\nu\rho^\nu \tag{81}$$

with the condition that the power series should terminate after a finite number of terms, i.e., n′ should be a finite integer.

Substituting the series form of $S(\varrho)$ from Equation 81 into Equation 79 and equating the coefficients of $\varrho^{\nu-1}$ from both sides of the equation, one obtains

$$\alpha_{\nu+1} = \alpha_\nu \frac{n-(\ell+\nu+1)}{\ell(\ell+1) - (\ell+\nu+1)(\ell+\nu+2)} \tag{82}$$

or

$$\alpha_{\nu+1} = \frac{\alpha_\nu\{\nu-(n-\ell-1)\}}{(\nu+1)(\nu+2\ell+2)} \tag{83}$$

If the series (Equation 81) has to terminate at n′, one demands that

$$\alpha_{\nu+1} = 0 \text{ for } \nu \geqq n' \tag{84}$$

which is possible from Equation 83 provided that

$$n = \ell + n' + 1 \tag{85}$$

or

$$n' = n - \ell - 1 \tag{86}$$

The condition (Equations 85 or 86) leads to the important result that, since $n' \geqslant 0$, n is an integer greater than zero. n and n′ are known as the principal quantum number and radial quantum number, respectively. If one knows n and ℓ, the radial quantum number n′ is redundant for specifying the eigenfunction and therefore n′ is frequently omitted.

The eigenfunction $\Psi(\vec{r})$ in Equation 23 can now be described explicitly in terms of the quantum numbers n, ℓ, and m as

$$\Psi_{n,\ell,m}(\vec{r}) = R_{n\ell}(r)\, Y_\ell^m(\theta,\phi) \tag{87}$$

The eigenvalues corresponding to these functions are obtained from Equations 75 and 76. Thus

$$E = -\frac{\alpha^2 \hbar^2}{8\mu} \tag{88}$$

where

$$\alpha = \frac{2\mu Z e^2}{4\pi\epsilon_o \hbar^2 n} \tag{89}$$

Explicitly, the eigenvalues are

$$E_n = -\frac{1}{2n^2} \frac{\mu Z^2 e^4}{(4\pi\epsilon_o)^2 \hbar^2} \tag{90}$$

where we have used the subscript n on the energy to show clearly that the energy depends on the quantum number n. The discreteness of the energy is manifested by the expression Equation 90. The energy levels of the hydrogenic atom have been depicted in Figure 5.

The energy level of the hydrogenic atom is multiple-degenerate since for every value of n there is a set of values of ℓ and m for which the energy remains the same. Explicitly, for a given n

$$\ell = n - \nu - 1 \tag{91}$$

and ν goes from 0 to $n - \ell - 1$ thereby giving the minimum and maximum values of ℓ as 0 and $n - 1$, respectively. Hence there are n values of ℓ, namely,

$$\ell = 0, 1, 2, \ldots (n - 1) \tag{92}$$

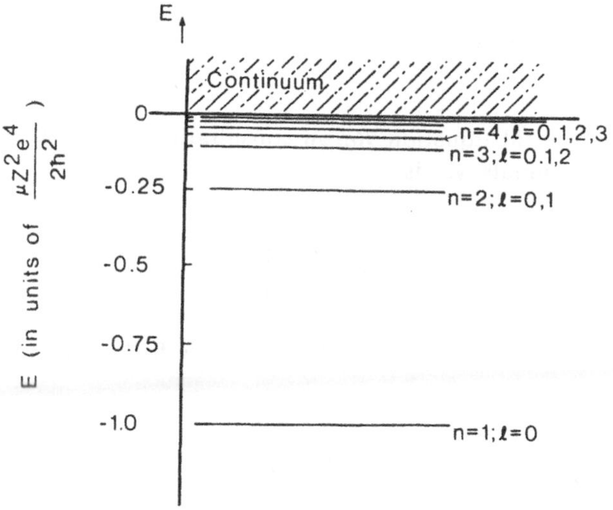

FIGURE 5. Hydrogenic atom energy level diagram.

for a definite value of the principal quantum number n. For each ℓ there are the following $(2\ell + 1)$ values of m

$$m = -\ell, -\ell + 1, \ldots -2, -1, 0, 1, 2, \ldots (+\ell) \quad (93)$$

Thus the degeneracy of an eigenstate prescribed by the principal quantum number n is

$$\sum_{\ell=0}^{n-1} (2\ell + 1) = n^2 \quad (94)$$

Some of these degeneracies can be removed if the perturbations from various interactions connected with spins or magnetic fields are present.

The polynomials,

$$\sum_{\nu=0}^{n'} \alpha_\nu \rho^\nu$$

with an appropriate normalization factor which terminates at $n' = n - \ell - 1$, are known as Laguerre polynomials and symbolized as $L_{n+\ell}^{2\ell+1}$. The solutions $S_{n\ell}(\varrho)$ of Equation 81 which satisfy Equation 79 are correspondingly Laguerre functions if the normalization constants are appropriately taken into account.

The radial functions can now be expressed as

$$R_{n\ell}(r) = S_{n\ell}(\rho) = A_{n\ell}\, e^{-\rho/2} \rho^\ell L_{n+\ell}^{2\ell+1}(\rho) \quad (95)$$

where we have incorporated the normalization constant $A_{n\ell}$ to satisfy the requirement

$$\int_0^\infty R^*_{n\ell}(r)\, R_{n\ell}(r)\, r^2\, dr = 1 \tag{96}$$

Evaluation of the integral Equation 96, on consultation of tables involving integrals of the Laguerre polynomials, yields

$$A_{n\ell} = \left\{ \left(\frac{2\mu Z e^2}{n\hbar^2} \right)^3 \frac{(n - \ell - 1)!}{2n\,[(n + 1)!]^3} \right\}^{1/2} \tag{97}$$

The expectation value of r for the ground state of the hydrogen atom ($Z = 1$) is obtained from

$$\bar{r} = \int_0^\infty R^*_{10}(r)\, R_{10}(r)\, r^2\, dr = \frac{\hbar^2}{\mu e^2} (4\pi\epsilon_0) \tag{98}$$

where

$$R_{10} = \left(\frac{1}{a_0} \right)^{3/2} 2 \exp(-r/a_0) \tag{99}$$

The quantity $\bar{r}$ for the ground state of the hydrogen atom is known as the Bohr radius a_o; i.e.,

$$a_0 = \frac{\hbar^2}{\mu e^2} (4\pi\epsilon_0) \tag{100}$$

Table 3 lists some of the radial functions $R_{n\ell}(r)$ for a hydrogenic atom. A graphical presentation of a few radial functions (r) is shown in Figure 6. In the spectroscopic notation $\ell = 0, 1, 2, 3, \ldots$, etc. states are known as s, p, d, f, . . . , etc., respectively. For the s-state for which $\ell = 0$, the radial function $R_{n\ell}$ is non-zero at $r = 0$; for all other values of ℓ, $R_{n\ell}$ vanishes at the origin, behaving as r_ℓ as r tends to zero. The form of the wave function near the origin, namely,

$$\text{Limit}_{(r \to 0)}\ R(r) \to 0 \quad (\ell \neq 0)$$

is of general validity for any potential V(r) which satisfies

$$\text{Limit}_{(r \to 0)}\ r^2 V(r) \to 0$$

This can be proved with the help of the Schrödinger Equation 25.

The radial functions $R_{n\ell}$ have the property that they have n′ number of nodes where

$$n' = n - \ell - 1$$

Table 3

LIST OF SOME OF THE NORMALIZED WAVE FUNCTIONS $R_{n\ell}(r)$ FOR THE HYDROGENIC ATOM; Ze IS THE CHARGE OF THE NUCLEUS AND

$$a_0 = \frac{\hbar^2 4\pi\epsilon_0}{\mu e^2}$$ WHERE μ = REDUCED MASS OF THE ELECTRON

$$R_{10}(r) = \left(\frac{Z}{a_0}\right)^{3/2} 2\,e^{-Zr/a_0}$$

$$R_{20}(r) = \left(\frac{Z}{2a_0}\right)^{3/2} 2\left(1 - \frac{Zr}{2a_0}\right) e^{-Zr/2a_0}$$

$$R_{21}(r) = \left(\frac{Z}{2a_0}\right)^{3/2} \frac{1}{\sqrt{3}} \frac{Zr}{a_0} e^{-Zr/2a_0}$$

$$R_{30}(r) = \left(\frac{Z}{3a_0}\right)^{3/2} 2\left(1 - \frac{2}{3}\frac{Zr}{a_0} + \frac{2}{27}\left(\frac{Zr}{a_0}\right)^2\right) e^{-Zr/3a_0}$$

$$R_{31}(r) = \left(\frac{Z}{3a_0}\right)^{3/2} \frac{4\sqrt{2}}{3} \frac{Zr}{a_0}\left(1 - \frac{1}{6}\frac{Zr}{a_0}\right) e^{-Zr/3a_0}$$

$$R_{32}(r) = \left(\frac{Z}{3a_0}\right)^{3/2} \frac{2\sqrt{2}}{27\sqrt{5}} \left(\frac{Zr}{a_0}\right)^2 e^{-Zr/3a_0}$$

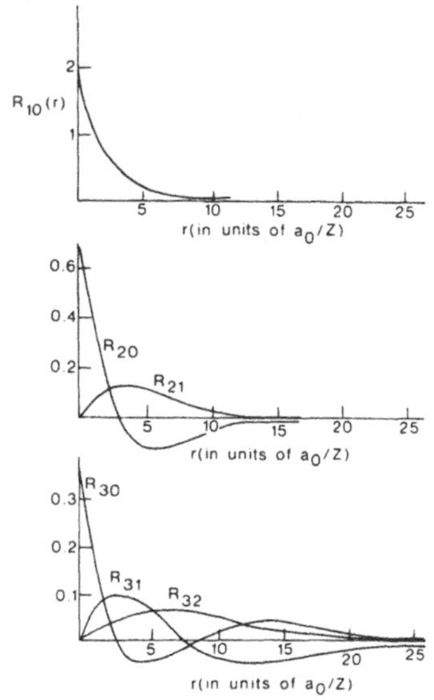

FIGURE 6. Plot of radial functions $R_{n\ell}(r)$ for a few values of n, ℓ quantum numbers; $a_0 = h^2 4\pi\varepsilon_0 / \mu e^2$.

F. Probability Density

The probability density, the probability per unit volume of finding the particle at a given location, is defined by

$$P(\vec{r},t) = \Psi^*(\vec{r},t)\,\Psi(\vec{r},t) \tag{101}$$

where $\Psi(r,t)$ is the wave function of the system. For the hydrogenic atom in the nth eigenstate it becomes

$$P(\vec{r},t) = \Psi^*_{n,\ell,m}(\vec{r})\, e^{iE_n t/\hbar}\, \Psi_{n,\ell,m}(\vec{r})\, e^{-iE_n t/\hbar}$$

$$P(\vec{r},t) = R^*_{n\ell}(r)\, \Theta^*_{\ell,m}(\theta)\, \Phi_m^*\, R_{n\ell}(r)\, \Theta_{\ell m}(\theta)\, \Phi_m(\phi)$$

$$= R^*_{n\ell}(r)\, Y_\ell^{m\,*}(\theta,\phi)\, R_{n\ell}(r)\, Y_\ell^m(\theta,\phi) \tag{102}$$

One notes that the probability density for the system in its eigenstate is independent of time. It is because of this that the eigenstates are termed stationary states. It is difficult to plot the three-dimensional probability density functions in two-dimensional space. To simplify the matter one usually plots the density functions by varying only one of the coordinates. If $P(\vec{r},t)$ is integrated over the volume contained between the spheres of radii r and r + dr, one obtains what is known as the radial probability density, that is,

$$P_{n\ell}(r)\,dr = R^*_{n\ell}(r)\, R_{n\ell}(r)\, r^2\, dr \tag{103}$$

This is the probability of finding the electron in the space between the radial coordinates r and r + dr. The factor r^2 in $P_{n\ell}(r)$ takes care of the volume lying between the spheres of radii r and r + dr. It is evident that the radial probability density does not depend on the quantum number m.

Some probability functions $P_{n\ell}(r)$ have been shown in Figure 7 which reveal that the probability density primarily depends on the principal quantum number n and only slightly on the quantum number ℓ. Also, there is a high probability of finding an electron in a restricted region of r space (see Figure 7), pointing out the fact that the electron is likely to be located in a so-called shell. The expectation value of r (for a given quantum state) is expressed as

$$\bar{r}_{n\ell} = \int \Psi^*(\vec{r},t)\, r\, \Psi(\vec{r},t)\, d^3r$$

$$= \int_0^\infty r\, P_{n\ell}(r)\, dr \tag{104}$$

which on evaluation yields

$$\bar{r}_{n\ell} = \frac{n^2 a_0}{Z}\left\{1 + \frac{1}{2}\left[1 - \frac{\ell(\ell+1)}{n^2}\right]\right\} \tag{105}$$

This formula demonstrates the strong dependence of $\bar{r}_{n\ell}$ on n^2 and the weak dependence on ℓ. In the Bohr model of the atom one obtains the expectation value of r as

$$r_{Bohr} = \frac{n^2 a_0}{Z} \tag{106}$$

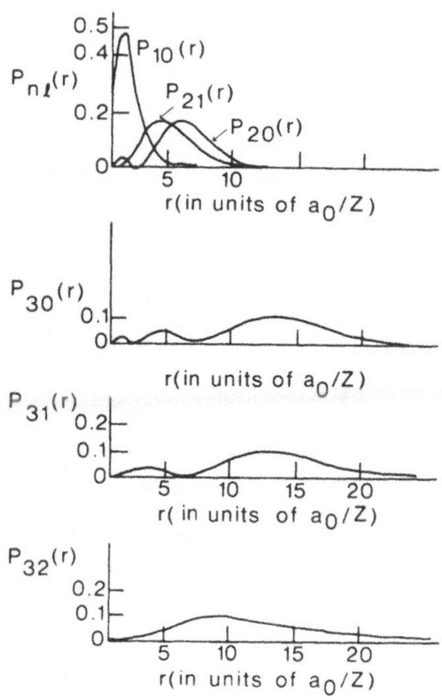

FIGURE 7. Some of the radial probability functions $P_{n\ell}$ (r).

which is very close to the above result (Equation 105) in spite of the fact that all the Bohr orbitals are circular. Notably, the expectation value of r increases with the principal quantum number n. The main reason for this is that the kinetic energy of the electron increases as n increases, with the result that the coordinate r of the electron increases with increasing n. In other words, the shells expand as n increases.

As concluded earlier, the hydrogenic wave function goes to zero as $r \rightarrow 0$ for all values of ℓ except for $\ell = 0$ so that the radial probability density functions have appreciable values near the origin only for the states which correspond to $\ell = 0$ (s-state). An important consequence of this is that the s-states interact more strongly with the nucleus than do the other states.

Next we turn to investigate the angular dependence of the probability density function $P(\vec{r},t)$ given in Equation 102. From the form of Φ_m (ϕ) (Equation 35) it is evident that P $(\vec{r},t)$ does not depend on the coordinate ϕ. The θ-dependent part contains the product Θ_ℓ^m (θ) * Θ_ℓ^m (θ), which is a θ-dependent modulating factor giving the directional effect on the radial density function P(r). For s, p, d, and f orbitals (i.e., ℓ = 0, 1, 2, and 3, respectively) we have shown the polar diagram in Figure 8 which gives the directional dependence of the one-electron atom probability density functions $\Theta_\ell^m{}^*(\theta)$ $(\Theta_\ell^m$ (θ). This figure is helpful in visualizing the three-dimensional characteristics of the probability density distribution function provided that one rotates the diagrams about the z-axis by 360° to cover the complete range of variation of the azimuthal angle ϕ. We remark that for $\ell = 0$ (the s-state),

$$\Psi^*_{n\ell m}(\vec{r}) \cdot \Psi_{n\ell m}(\vec{r})$$

does not depend on θ and ϕ and consequently, the probability density is spherically

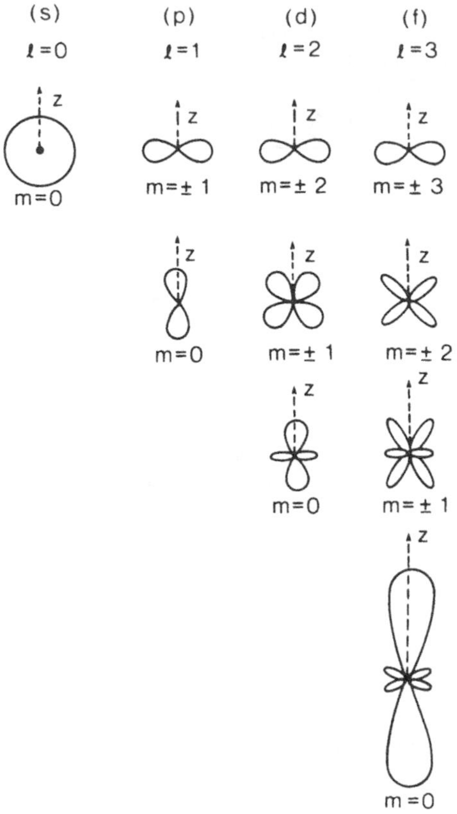

FIGURE 8. Electron probability density distribution $\Theta_{\ell}^{m_\ell *}\Theta_{\ell}^{m_\ell}$ depicting the directional dependence.

symmetric. For a given ℓ, the probability density associated with the m = o state is more pronounced on the z-axis; as $|m|$ increases in value to ℓ, the probability density shifts more and more towards the plane perpendicular to the z-axis. Beautifully shaded pictures depicting the three-dimensional appearance of the probability density functions have been given by Leighton[3] for various states of a one-electron atom. The fact that a bound particle possesses standing waves with fixed nodes is clear from the diagrams in Figure 8 since they generate nodal surfaces in three-dimensional diagrams.

In actual practice it is not possible to observe experimentally any of the probability density distribution functions of Figure 8 for a free atom (i.e., when no external electric or magnetic fields are present). It happens that, because of the degeneracy of the states (of a given n) with respect to the ℓ values, the probability density averages out to yield a spherically symmetric function. A similar situation also exists for each subset of states for a given n and ℓ if one includes all possible values of m to evaluate the average value of the probability densities. As an example, $\Psi^*_{200}\Psi_{200}$ is spherically symmetric and so also $\Psi^*_{21-1}\Psi_{21-1} + \Psi^*_{210}\Psi_{210} + \Psi^*_{211}\Psi_{211}$.

The situation is different when the z-direction is a preferred direction, as is the case when an electric or magnetic field is applied to an atom. In such a circumstance the quantum states are no longer degenerate and it is possible to obtain information by measurements about the probability density of atoms in a particular quantum state.

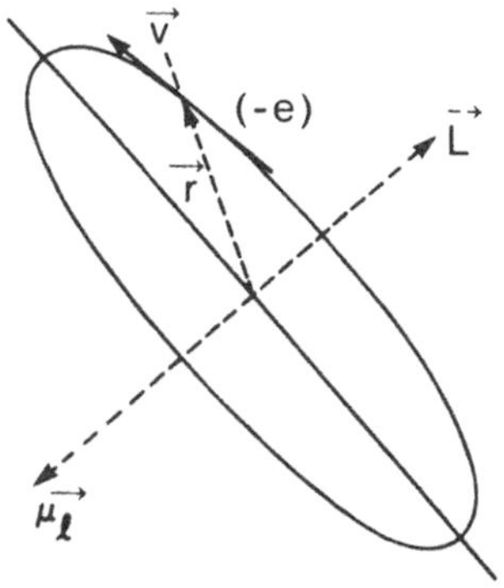

FIGURE 9. Representation of an electron moving with velocity $\vec{v}$ in a Bohr orbit. $\vec{L}$ and $\vec{\mu_\ell}$ are the orbital angular momentum and the associated orbital magnetic moment, respectively. $\vec{\mu_\ell}$ opposes $\vec{L}$ in direction.

G. Relation Between Orbital Magnetic Moment and Orbital Angular Momentum

To keep our treatment simple and clear, we employ here a combination of classical quantum physics (such as the Bohr model) and modern quantum mechanics to develop a relation between $\vec{\mu\ell}$, the orbital magnetic dipole moment and $\vec{L}$, the orbital angular momentum. A full quantum mechanical approach will require advanced knowledge of electromagnetic theory which we shall not assume at present. Further, the results obtained by this procedure are found to agree perfectly with those obtained by complete quantum mechanical treatment.

For the purpose of deriving a relation between $\vec{\mu\ell}$ and $\vec{L}$ we consider that an electron of mass m and charge (−e) moves with velocity $\vec{v}$ in a Bohr orbit of radius r (see Figure 9). The electron charge circulating in the circular orbit constitutes an electric current given by

$$i = \frac{\text{charge}}{\text{orbital period}} = \frac{(-e)v}{2\pi r} \qquad (107)$$

From classical electrodynamics the current i in the loop of area πr^2 is equivalent to a magnetic dipole with moment

$$\mu_\ell = i\pi r^2 \qquad (108)$$

in the direction perpendicular to the plane of orbit, as shown in Figure 9. The orbital angular momentum of the electron is

$$\vec{L} = \vec{r} \times m\vec{v} \qquad (109)$$

The directions of $\vec{r}$, $\vec{v}$, and $\vec{L}$ are illustrated in Figure 9. Combining Equations 107 to 109 we have the vector relation

$$\vec{\mu_\ell} = -\frac{e}{2m}\vec{L} \qquad (110)$$

Rewriting the above relation in the usual way,

$$\vec{\mu}_\ell = -g_\ell \frac{\mu_B}{\hbar} \vec{L} \tag{111}$$

where μ_B is the Bohr magneton,

$$\mu_B = \frac{e\hbar}{2m} = 0.927 \times 10^{-23} \text{ amp-m}^2 \tag{112}$$

and g_ℓ the orbital g-factor,

$$g_\ell = 1 \tag{113}$$

The g-factor has been introduced here in order to retain the similarity between the expression in Equation 111 and the expressions for $\vec{\mu}_\ell$ one obtains in complicated cases of the motion of the electrons where the g-factor is not actually unity. Though the relation Equation 111 has been derived here assuming a circular orbit, it can be proved that it does not depend on the shape of the orbit; in other words, a similar relation can be obtained for an elliptical orbit. Consequently the ratio of μ_ℓ to L is independent of the details of the mechanical theory used to derive it. A more sophisticated treatment using complete quantum mechanical theory produces the same result. In the quantum mechanical picture one then writes the magnitude of the orbital magnetic dipole moment as

$$\mu_\ell = g_\ell \mu_B \sqrt{\ell(\ell+1)} \tag{114}$$

since, in quantum mechanical theory, the magnitude of the orbital angular momentum is $\hbar\sqrt{\ell(\ell+1)}$. Analogously, the z-component of the orbital angular magnetic moment has the expectation value

$$\mu_{\ell_z} = -g_\ell \mu_B m_\ell \tag{115}$$

where m_ℓ is the magnetic quantum number associated with ℓ. Equation 115 has been obtained by substituting the expectation value $m_\ell\hbar$ for L_z which appears if the z-component of $\vec{\mu}_\ell$ in Equation 111 is considered.

In a magnetic field $\vec{B}$, following the principles of elementary electromagnetic theory, the dipole moment $\vec{\mu}_\ell$ experiences a torque $\vec{\tau}$ given by

$$\vec{\tau} = \vec{\mu}_\ell \times \vec{B} \tag{116}$$

This torque causes the dipole moment $\vec{\mu}_\ell$ to align with the magnetic field $\vec{B}$. The associated potential energy of orientation is

$$\Delta E = -\vec{\mu}_\ell \cdot \vec{B} \tag{117}$$

One needs to supply energy equivalent to $2\mu_\ell\vec{B}$ to turn the dipole μ_ℓ to align it antiparallel to $\vec{B}$ if it is initially aligned parallel to $\vec{B}$. For instance, the energy required is about 2×10^{-4} eV for μ_ℓ equal to about 1 μ_B in the field of 2 T. On the other hand, if the dipole is initially aligned antiparallel to $\vec{B}$, it can be brought into the configuration of parallel alignment only when there is some way of getting rid of this much change in energy.

It happens that when there is no way to releasing the energy the magnetic dipole moment follows the basic classical law which states that the rate of change of angular momentum is equal to the impressed torque, i.e.,

$$\frac{d\vec{L}}{dt} = \vec{\tau} \tag{118}$$

Consequently the angular momentum precesses around $\vec{B}$, keeping the energy constant. The angle between μ_ℓ and $\vec{B}$ remains constant and the frequency of μ_ℓ about $\vec{B}$ is

$$\vec{\omega} = \frac{g_\ell \mu_B}{\hbar} \vec{B} \tag{119}$$

which is known as the Larmor frequency, and the precession of $\vec{\mu}_\ell$ about $\vec{B}$ is referred to as the Larmor precession.

H. Electron Spin and the Stern-Gerlach Experiment

Stern and Gerlach[23] performed experiments to measure the magnetic dipole moment of neutral silver atoms. They used a beam of these atoms and passed it through a nonuniform magnetic field, as shown in Figure 10. If the magnetic dipole moment of the atoms is μ_l, the nonhomogeneous magnetic field produces an average transverse force on each atom of magnitude given by

$$\bar{F}_z = \mu_{lz} \left[\frac{\partial B_z}{\partial z}\right] \tag{120}$$

where $\partial B_z/\partial z$ is the gradient of the field in the z-direction and μ_{lz} is the component of the vector μ_l along the field-gradient direction. According to the classical treatment, the atomic beam would spread out because of the random orientations of the dipole moments of the atoms. In the Stern-Gerlach experiment the silver atomic beam was actually split into two discrete components, furnishing evidence of two possible values of μ_{lz}. The force exerted on the atoms is proportional to μ_{lz} and, in accordance with quantum mechanics, μ_{l_z} can have the discretely quantized values

$$\mu_{lz} = -g_\ell \mu_B m_\ell \tag{121}$$

with

$$m_\ell = \ell, \ell - 1, \ell - 2, \ldots, (-\ell + 1), (-\ell) \tag{122}$$

Similar experiments on different species of atoms also showed that the deflected atomic beam was separated into two or more discrete components. The results demonstrate directly the quantization of the z-component of the magnetic dipole moment of atoms in space, giving rise to the phenomenon known as space quantization.

If ℓ is an integer there are $(2\ell + 1)$ number of m_l values, and then the number of split components of the beam is expected to be odd in number. However, Stern-Gerlach's experiment on neutral silver atoms actually produced only two components,

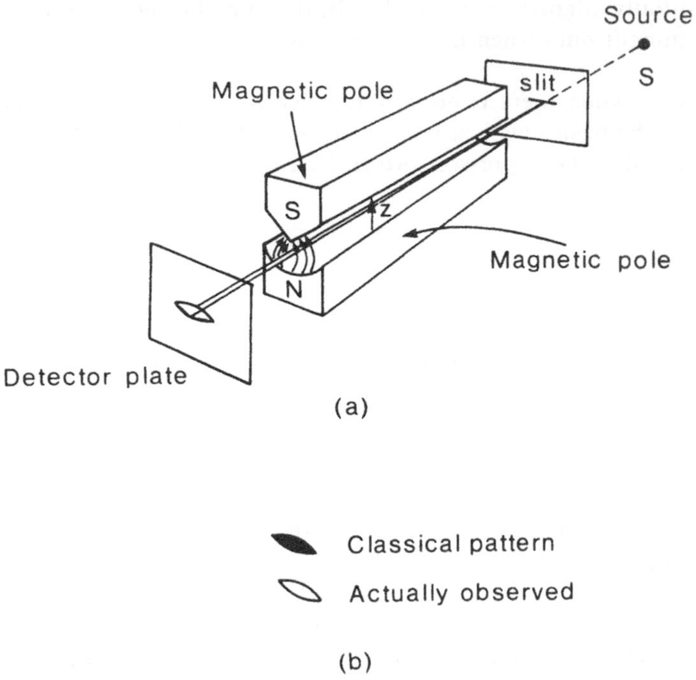

FIGURE 10. Illustration of Stern-Gerlach experimental apparatus. (a) Beam of neutral atoms is directed along the sharp edge of a magnetic pole. The field lines have been shown at the ends of the pole pieces. The magnetic field increases in the z-direction as shown. The nonuniform magnetic field exerts a transverse force, thereby deflecting the atomic beam. (b) Comparison of classically predicted and actually observed patterns.

establishing that ℓ is not an integer but equal to ½. This can be understood only when one assumes that an electron has an intrinsic magnetic angular momentum $\vec{S}$, termed spin, with only two possible z-components of S. Explicitly, one assumes that the magnitude of $\vec{S}$ and the z-component S_z of the spin angular momentum are given by (in analogy with the orbital angular momentum)

$$S = \sqrt{s(s+1)}\,\hbar \tag{123}$$

$$S_z = \hbar m_s \tag{124}$$

with

$$s = \frac{1}{2} \tag{125}$$

and

$$m_s = -\frac{1}{2}, \frac{1}{2} \tag{126}$$

The spin magnetic dipole moment $\vec{\mu}_s$ associated with $\vec{S}$ is

$$\vec{\mu}_s = -\frac{g_s \mu_B}{\hbar} \vec{S} \tag{127}$$

where μ_B is the Bohr magneton as in Equation 112.

The z-component of μ_s is given by,

$$\mu_{sz} = -g_s \mu_B m_s \tag{128}$$

where g_s is the spin g-factor.

By measuring the splitting of the beam on the detector plate and knowing the magnetic field gradient $\partial B_z / \partial z$, experimental analysis gave the result

$$g_s m_s = \pm 1 \tag{129}$$

implying that

$$g_s = 2 \tag{130}$$

These results were also supported by the experiments performed by Phipps and Taylor[24] who used a beam of hydrogen atoms and a Stern-Gerlach-type apparatus to deflect the beam. A very accurate value of the spin g-factor has now been determined from the recent spectroscopic measurements by Lamb and Rutherford[25] which gave

$$g_s = 2.00232$$

Though Compton[26] had speculated in 1921 in his article on the scattering of X-rays by atoms that the spinning of an electron is analogous to a tiny gyroscope, it is Goudsmit and Uhlenbeck[27] who are given credit for introducing the electron spin to explain the fine structure (closely spaced pair of lines in the optical spectra) of hydrogen and alkali atoms. Previously Sommerfeld had tried to explain the fine structure of atoms by considering the effect (about $1{:}10^4$ on the energy levels) of the relativistic variation of the electron mass with velocity on the energy levels obtained in the Bohr model. Sommerfeld's explanation was satisfactory for the hydrogen atoms. However, for the alkali atoms the electron responsible for the optical spectra is expected to have a large Bohr radius and small velocity. This would give a small relativistic correction in contrast to the experimentally observed value which is very much larger. This led Goudsmit and Uhlenbeck[27] to propose an intrinsic angular momentum (spin) and the associated magnetic dipole moment ($\vec{\mu}_s$) with the z-component of the spin assuming only two values $\frac{1}{2}\hbar$ and $-\frac{1}{2}\hbar$, corresponding to the spin being "up" or "down". The interaction between the spin magnetic dipole moment $\vec{\mu}_s$ and the internal magnetic field experienced by the electron was then used to explain the required fine splitting.

Thus the description of an electron requires a set of four quantum numbers: n, ℓ, m_ℓ, and m_s. The first three quantum numbers arise because of its three independent spatial coordinates and the fourth one arises because it is necessary to describe the orientation in space of its spin relative to some z-axis.

Schrödinger's quantum mechanics does not predict spin; it must be introduced externally. This is because the Schrödinger theory is a nonrelativistic theory. Dirac[28] developed a relativistic theory using the same postulates as the Schrödinger theory and predicted that the spin carries an intrinsic angular momentum $s = \frac{1}{2}\hbar$, thereby putting the electron spin on a firm theoretical footing and revealing that the electron spin is actually connected with the theory of relativity.

As a simple example it is worth considering a hydrogen atom in its ground state in an external magnetic field. The electron being in the 1s state possesses an orbital angular momentum equal to zero but it has a spin $\vec{s}$. The external magnetic field $\vec{B}$ interacts with the spin magnetic moment of the electron giving rise to an orientational potential energy

$$\begin{aligned} \Delta E &= -\vec{\mu}_s \cdot \vec{B} \\ &= -\mu_{sz} B \\ &= g_s \mu_B B\, m_s \\ &= \pm \frac{1}{2} g_s \mu_B B \end{aligned} \tag{131}$$

where the z-axis is along the magnetic field. The above interaction is known as the Zeeman interaction and gives rise to the splitting of the energy levels of the electron in a magnetic field — the Zeeman effect.[29]

I. The Spin-Orbit Interaction

The spin-orbit interaction arises from the interaction between the magnetic dipole moment associated with the electron spin and the internal magnetic field connected with the orbital angular momentum of the electron. It is responsible for what is known as the fine structure of the excited states of atoms. The spin-orbit interaction is relatively weak in one-electron atoms whereas in multielectron atoms it is reasonably strong because of the strong internal magnetic fields. In the case of nucleons in the nuclei, the spin-orbit interaction is very strong and dominates the properties of the nuclei.

In order to understand the origin of the spin-orbit interaction we follow a simple picture as illustrated in Figure 11. The electron of the atom experiences not only the electric field $\vec{E}$ due to the nucleus, but also the relative motion of the nucleus. In the frame of reference fixed at the electron, the nucleus of charge Ze moves around the electron and produces a magnetic field $\vec{B}$ at the electron. If the electron velocity with respect to the nucleus is $\vec{v}$, the velocity of the nucleus with respect to the electron is $-\vec{v}$. This constitutes a current element $\vec{j}$ given by

$$\vec{j} = -Ze\vec{v} \tag{132}$$

(Note that $\vec{j}$ is not the current density but is actually the current density multiplied by the volume element.)

In accordance with Ampere's law the magnetic field at the location of the electron is

$$\vec{B} = \frac{\mu_o}{4\pi} \frac{\vec{j} \times \vec{r}}{r^3} = \frac{-Ze\mu_o}{4\pi} \frac{\vec{v} \times \vec{r}}{r^3} \tag{133}$$

Also, the electric field at the electron due to the nuclear charge Ze is, from Coulomb's law,

$$\vec{E} = \frac{Ze}{4\pi\epsilon_o} \frac{\vec{r}}{r^3} \tag{134}$$

Combining Equations 133 and 134 we have

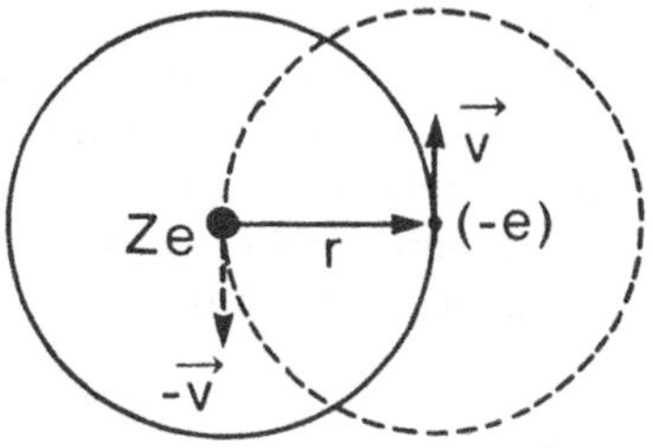

FIGURE 11. The motion of the electron as seen by the nucleus has been depicted by the solid circle. In the same situation, the motion of the nucleus as seen by the electron has been shown by the dotted circle. Note that the nucleus moves with velocity $(-\vec{v})$ relative to the electron if the velocity of the electron is $\vec{v}$ in the frame of reference of the nucleus. Because of this the magnetic field at the location of the electron is in the direction out of the page.

$$\vec{B} = \frac{1}{c^2}\,\vec{v} \times \vec{E} \tag{135}$$

where c is the velocity of light

$$c = (1/\sqrt{\mu_0 \epsilon_0})$$

Though we have derived Equation 135 from a simple picture it is in fact very general and can be derived from relativistic considerations. Also, it is interesting to note from Equation 135 that the electron experiences a magnetic field as long as it moves with some velocity in the presence of an electric field.

Knowing $\vec{B}$, one writes the orientational potential energy of the magnetic spin dipole moment $\vec{\mu}_s$ of the electron as

$$\Delta E = -\vec{\mu}_s \cdot \vec{B} \tag{136}$$

with

$$\vec{\mu}_s = -g_s \mu_B \frac{\vec{S}}{\hbar} \tag{137}$$

where g_s, as before, is the g-factor for the spin S of the electron and μ_B is the Bohr magneton $e\hbar/2m$.

Thus,

$$\Delta E = \frac{g_s \mu_B}{\hbar}\,\vec{S} \cdot \vec{B} \tag{138}$$

Equation 138 refers to the frame in which the electron is at rest and the nucleus is in motion. However, we are interested in the energy in the normal frame of reference in which the nucleus is at rest. Thus we transform the velocity back to the nuclear frame of reference by means of the relativistic transformations. This gives rise to the effect known as the Thomas precession[30] and effectively reduces the orientational potential energy in Equation 138 by a factor, 2. Thus,

$$\Delta E = \frac{1}{2} \frac{g_s \mu_B}{\hbar} \vec{S} \cdot \vec{B} \tag{139}$$

Now, the relation between the potential V(r) and $\vec{E}$ is

$$(-e)\vec{E} = \left(\frac{-dV(r)}{dr}\right) \frac{\vec{r}}{r} \tag{140}$$

Next, the use of the orbital angular momentum

$$\vec{L} = \vec{r} \times \vec{p} = \vec{r} \times m\vec{v}$$

transforms Equation 139 into the form

$$\Delta E = \frac{g_s \mu_B}{2\, emc^2 \hbar} \frac{1}{r} \frac{dV(r)}{dr} \vec{S} \cdot \vec{L}$$

or

$$\Delta E = \frac{1}{2\, m^2\, c^2} \frac{1}{r} \frac{dV(r)}{dr} \vec{S} \cdot \vec{L} \tag{141}$$

This is the spin-orbit interaction and was first derived by Thomas in 1926. Though we have used a mixture of the Bohr model, Schrödinger theory and relativistic transformations, the result is the same as one gets from the relativistic quantum theory of Dirac. The spin-orbit interaction is of central importance in atoms, molecules, solids, and nuclei.

An idea of the approximate spin-orbit interaction energy can be obtained by evaluating Equation 141. For the hydrogen atom in the $n = 1$, $\ell = 0$ (1s) state, the spin-orbit interaction energy is zero. However, for the $n = 2$, $\ell = 1$ (2p) state, the direct evaluation yields

$$[\Delta E] \sim 10^{-4} \text{ eV} \tag{142}$$

This is in reasonable agreement with the experimental observation of the fine structure splitting of the hydrogen energy level associated with the 2p level. The magnitude of the magnetic field in the present case is of the order of 1 T, a large value (being almost equal to the saturation magnetic field in an electromagnet), which is not surprising since the electron moves with a high velocity in the strong electric field created by the nucleus.

J. Addition of Angular Momenta to Form Total Angular Momentum

We have seen that the orbital angular momentum $\vec{L}$ precesses about the z-axis in such a way that the magnitude of $\vec{L}$ and its z-component, L_z, remain constant in the absence of an external perturbing field. Similarly, if no spin-orbit interaction exists, the spin angular momentum would also precess independently about the z-axis maintaining the magnitude of $\vec{S}$ and its z-component S_z constant. In that case one describes the electron wave function by specifying the four quantum numbers ℓ, m_ℓ, s, and m_s.

We know that an atomic electron experiences a magnetic field that is proportional to its orbital angular momentum $\vec{L}$. This field interacts with the electron spin magnetic moment $\vec{S}$, thereby producing a coupling between the vectors $\vec{L}$ and $\vec{S}$ (see Equation 141). With this interaction neither the magnitude of $\vec{L}$ nor that of $\vec{S}$ is changed, which is obvious since the situation is analogous to that of the Larmor precession. However, the orientations of $\vec{L}$ and $\vec{S}$ are dependent on each other because of the coupling between them due to spin-orbit interaction. The quantum mechanical angular momentum conservation law requires that the total angular momentum

$$\vec{J} = \vec{L} + \vec{S} \tag{143}$$

must be conserved. That is, the magnitude of $\vec{J}$ and its z-component J_z must remain constant. To satisfy this requirement the $\vec{L}$ and $\vec{S}$ vectors precess about the vector $\vec{J}$. The situation has been depicted in Figure 12. This method of coupling the angular momenta is a useful one, particularly in case of multielectron atoms where one needs to combine many orbital and spin angular momenta to form the conserved total angular momentum. The method is known as Russell-Saunders coupling[31,32] or LS coupling.

Treating the total angular momentum, $\vec{J}$ and its z-component J_z, in the same way as the orbital angular momentum, $\vec{L}$ and its z-component L_z (Section II.D) and assigning the corresponding quantum numbers j and m_j, we write the quantization rules

$$J = \sqrt{j(j+1)}\ \hbar \tag{144}$$

$$J_z = m_j\hbar \tag{145}$$

where

$$m_j = +j, j-1, \ldots, (-j+1), (-j) \tag{146}$$

Taking the z-component of Equation 143, we have

$$J_z = L_z + S_z \tag{147}$$

which, in terms of the quantum numbers m_j, m_ℓ, and m_s becomes

$$m_j = m_\ell + m_s \tag{148}$$

Now we encounter the problem of determining the maximum and minimum allowed values of the quantum number j appearing in Equation 144. These values can easily be obtained by vector addition of $\vec{L}$ and $\vec{S}$ as explained in Figure 13 which gives

$$j_{max} = \ell + s \tag{149}$$

and

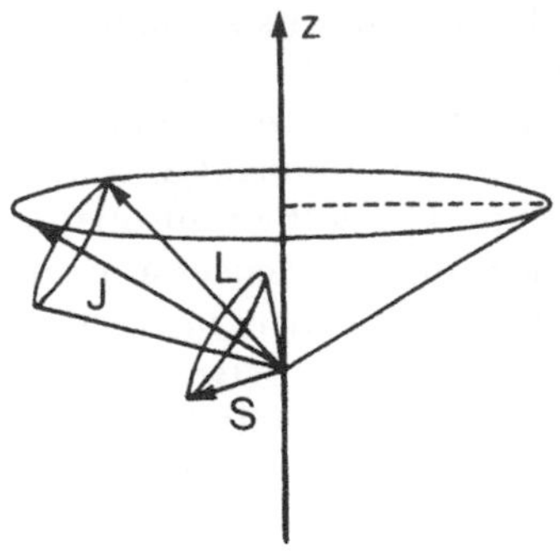

FIGURE 12. Representation of the total angular momentum $\vec{J}$, which is the vector sum of $\vec{L}$ and $\vec{S}$. The vector $\vec{J}$ precesses about the z-axis whereas $\vec{L}$ and $\vec{S}$ precess about $\vec{J}$.

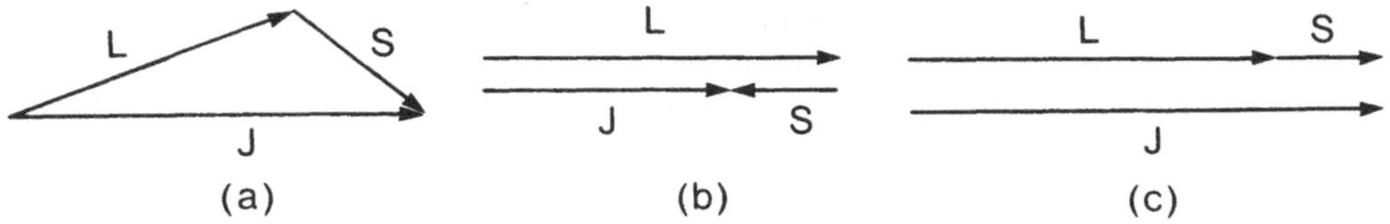

FIGURE 13. Vector addition of $\vec{L}$ and $\vec{S}$ to form $\vec{J}$. (a) A general diagram showing $\vec{J} = \vec{L} + \vec{S}$. (b) The minimum value of $\vec{J}$, which is the absolute difference of the magnitudes of $\vec{L}$ and $\vec{S}$. (c) The maximum value of $\vec{J}$, which is the sum of the magnitudes of $\vec{L}$ and $\vec{S}$.

$$j_{min} = |\ell - s| \tag{150}$$

The intermediate values of j differ by unity. Thus,

$$j = \ell + s, (\ell + s - 1), \ldots |(\ell - s)| \tag{151}$$

In the present case of a single electron, s = ½ and therefore,

$$j = (\ell + \tfrac{1}{2}), |\ell - \tfrac{1}{2}| \tag{152}$$

The maximum value of j is consistent with the expressions of Equations 146 and 148 since $(m_j)_{max}$ from Equation 148 is

$$(m_j)_{max} = \ell + s \tag{153}$$

and $(m_j)_{max}$ from Equation 146 is j.

It is apparent from Relation 152 that for $\ell = 0$, j has only one value, viz.,

$$j = \tfrac{1}{2} \tag{154}$$

For $\ell = 1$ and s = ½, one notes that j = 3/2 and ½; and for $\ell = 2$ and s = ½, j = 5/2 and 3/2, which follow directly for Equation 152.

Equation 151 is a general expression which is useful for obtaining the total quantum number corresponding to the combination of any two angular momenta — whether they are both orbital or both spin, or mixed.

K. Splitting of Hydrogen Energy Levels by the Spin-Orbit Interaction

We have already derived an expression (see Equation 141) for the interaction between the spin of an electron and its orbital motion. Here we shall show that this interaction explains the detailed structure of the energy levels of the hydrogen atom. Further use of it will become clear when we treat the structure of multielectron atoms.

In the absence of the spin-orbit interaction (see Equation 90) we have found the energy levels E_n of the hydrogen atom by solving the time-independent Schrödinger equation. When the spin-orbit interaction is present, the Hamiltonian of the hydrogen atom is changed to include the additional interaction as given by Equation 141. The expectation value of the interaction can be obtained easily if one notes that since $\vec{J} = \vec{L} + \vec{S}$,

$$J^2 = L^2 + S^2 + 2\vec{S}\cdot\vec{L} \tag{155}$$

and hence,

$$S\cdot L = \tfrac{1}{2}(J^2 - L^2 - S^2) \tag{156}$$

Rewriting Equation 141 in view of Equation 156:

$$\Delta E = \frac{1}{4m^2c^2}\,\frac{1}{r}\,\frac{dV(r)}{dr}\,(J^2 - L^2 - S^2) \tag{157}$$

One needs to evaluate the expectation value of ΔE in the above equation for a quantum state characterized by the quantum numbers ℓ, s, and j. Replacing the expectation values of the operators (see Equations 70, 123, and 144), Equation 157 yields the spin-orbit interaction energy

$$\Delta E_{S-O} = \zeta_{n\ell}\,\frac{1}{2}\,[j(j+1) - \ell(\ell+1) - s(s+1)] \tag{158a}$$

where $\zeta_{n\ell}$ is the spin-orbit coupling constant for the state $n\ell$,

$$\zeta_{n\ell} = \frac{\hbar^2}{2m^2c^2}\left\langle \frac{1}{r}\,\frac{dV}{dr}\right\rangle_{n\ell} \tag{158b}$$

and <(1/r)(dV/dr)> is the expectation value of (1/r)(dV/dr) for the electron in the quantum state $\psi_{n\ell m_\ell}$ That is,

$$\left\langle \frac{1}{r}\,\frac{dV}{dr}\right\rangle_{n\ell} = \int \psi^*_{n\ell m_\ell}\left(\frac{1}{r}\,\frac{dV}{dr}\right)\psi_{n\ell m_\ell}\, d^3r \tag{159}$$

According to Equation 158a it is obvious that there is no effect on the energy of the hydrogen atom in the s-state (corresponding to which $\ell = 0$), since then $j = \tfrac{1}{2}$ and $\Delta E = 0$. However, for states for which $\ell \neq 0$ (p,d,f, etc. states) the spin-orbit interaction splits the energy levels in accordance with Equation 158a. For instance the energy level corresponding to the n = a state of the hydrogen atom undergoes splitting, which can be calculated by considering the allowed azimuthal quantum numbers ℓ = 0 and ℓ = 1 for this case. The two values of ℓ can be handled separately. For ℓ = 0 we have ΔE = 0 as we have seen above. However, for ℓ = 1 there are two values of j, namely ½ and 3/2. These values yield from Equation 158a, for n,ℓ,j = 2,p,½,

$$\Delta E_{S-O} = \zeta_{2p} \quad (\text{for } j = \frac{1}{2}, \ \ell = 1) \tag{160}$$

and, for n,ℓ,j = 2,p,3/2,

$$\Delta E_{S-O} = \frac{1}{2}\zeta_{2p} \quad (\text{for } j = \frac{3}{2}, \ \ell = 1) \tag{161}$$

The spin-orbit coupling constant for the hydrogen atom is positive and its order of magnitude comes out to be 10^{-4}eV. Accordingly, the 2p energy level of the hydrogen atom is split by the spin-orbit interaction alone into three lines for which the quantum numbers n,ℓ,j are 2,p,½, 2, s, ½, and 2,p,3/2 in increasing order of energy. It may be concluded that for the given values of ℓ and s, the energy level with higher values of j lies still higher.

For the one-electron atom, the expectation value in Equation 159 can easily be carried out in a straightforward way. Now, for the potential seen by the electron,

$$\left\langle \frac{1}{r} \frac{dV}{dr} \right\rangle = \frac{Ze^2}{4\pi\epsilon_o} \left\langle \frac{1}{r^3} \right\rangle_{n\ell} \tag{162}$$

where $\langle 1/r^3 \rangle_{n\ell}$ is the expectation value of $1/r^3$ over the unperturbed function $\psi_{n\ell m_\ell}$. Using the generating function representation[17,33] or otherwise one obtains,

$$\left\langle \frac{1}{r^3} \right\rangle_{n\ell} = \frac{2Z^3}{a_o^3 n^3 \ell(2\ell+1)(\ell+1)} \tag{163}$$

Thus, in general, for one-electron atoms

$$\langle \Delta E \rangle_{S-O} = \frac{Z^4 e^2 \hbar^2 \frac{1}{2} [j(j+1) - \ell(\ell+1) - s(s+1)]}{(4\pi\epsilon_o)\, m^2 c^2 a_o^3 \ell(2\ell+1)(\ell+1) n^3} \tag{164}$$

The other effect which is of comparable importance is the relativistic effect due to the (relativistic) dependence of the electron mass on its velocity, which gives rise to a shift in the energy of the same order of magnitude as the spin-orbit interaction. A refined treatment should include in the Schrödinger theory the correction due to not only the spin-orbit interaction, but also due to the relativistic dependence of the electron mass on the velocity. It is possible to obtain the corrections due to the relativistic mass dependence on the velocity as follows, without going into the details of quantum electrodynamics. The relativistic Hamiltonian operator of a particle of rest mass m and momentum $\vec{p}$ in a potential field V is

$$\mathcal{H} = \sqrt{(p^2c^2 + m^2c^4)} - mc^2 + V \tag{165}$$

In the limit of $p << mc$, the Hamiltonian can be expanded to yield

$$\mathcal{H} = \frac{p^2}{2m} - \frac{p^4}{8m^3c^2} + \ldots + V \tag{166}$$

The terms $p^2/2m$ and V have already been considered in Schrödinger's theory, so the largest correction term due to the relativistic effect is $(-p^4/8m^3c^2)$. This term acts as a perturbation on the energy levels obtained by Schrödinger's theory. The expectation value of this shift over the unperturbed functions $\psi_{n\ell m_\ell}$ is

$$\langle \Delta E \rangle_r = -\int \Psi_{n\ell m_\ell} \left(\frac{p^4}{8m^3c^2}\right) \Psi_{n\ell m_\ell} \, d^3r \tag{167}$$

which can be evaluated either directly or by making use of the relation

$$p^4/8m^3c^2 = \frac{T^2}{2mc^2} = (E_n - V)^2/2mc^2$$

or

$$\langle \Delta E \rangle_r = -\frac{1}{2mc^2} [E_n^2 - 2E_n \langle V \rangle + \langle V^2 \rangle] \tag{168}$$

One obtains, for a single-electron atom

$$\langle \Delta E \rangle_r = -\frac{Z^2\alpha^2}{4n^2} E_n \left[3 - \frac{4n}{\left(\ell + \frac{1}{2}\right)}\right] \tag{169}$$

where the subscript r stands for the relativistic contribution, α is the fine structure constant

$$\alpha = \frac{e^2}{4\pi\epsilon_o \hbar c} \simeq \frac{1}{137} \tag{170}$$

and E_n is given by Equation 90.

The combined effect of the spin-orbit interaction and the relativistic correction gives the refined one-electron atomic energy levels E'_n,

$$E'_n = E_n + \langle \Delta E \rangle_{S-O} + \langle \Delta E \rangle_r$$

$$= -\frac{\mu Z^2 e^4}{(4\pi\epsilon_o)^2 2\hbar^2} \frac{1}{n^2} \left[1 + \frac{\alpha^2}{n} \left\{\frac{1}{j + \frac{1}{2}} - \frac{3}{4n}\right\}\right] \tag{171}$$

This is exactly the same expression as derived by Dirac[28] using relativistic quantum mechanics, which is a complete treatment incorporating all effects of relativity (recalling that the spin-orbit interaction is also a relativistic effect). The result derived by Sommerfeld,[34] by including the relativistic effect of the variation of mass with velocity on the Bohr energy levels, was essentially the same as in Equation 171. This coincidence

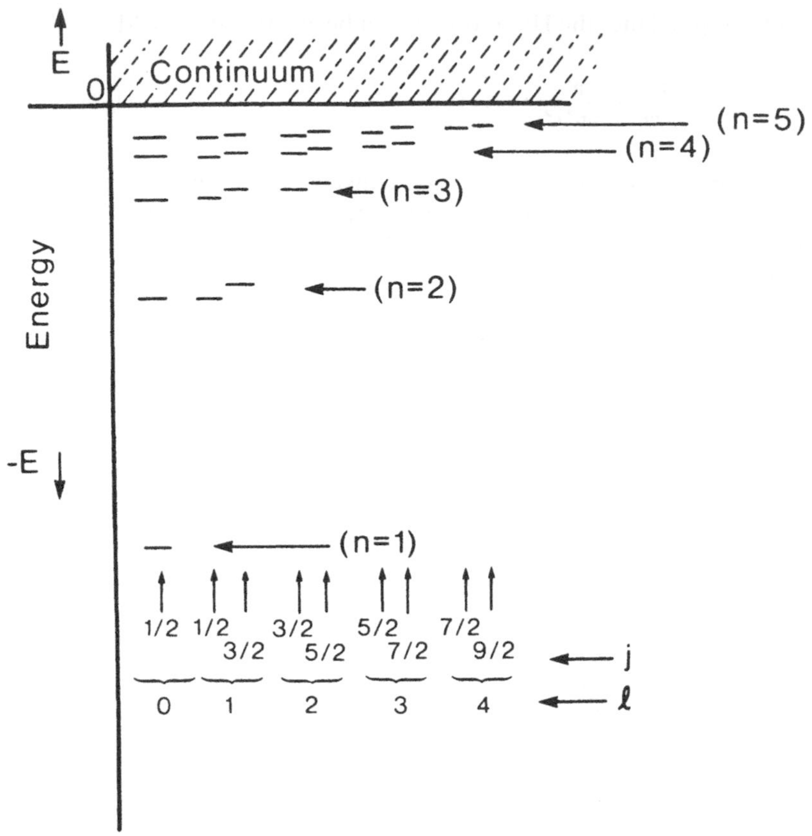

FIGURE 14. Hydrogen energy levels from the Dirac theory which includes all the relativistic and spin effects.

occurred because of the cancellation of errors introduced in using classical mechanics for the evaluation of the relativistic energy shift and in ignoring the spin-orbit interaction.

The optical spectroscopic measurements of the energy levels of the hydrogen atom are in very good agreement with the energy levels from Equation 171 obtained by both Sommerfeld and Dirac. Figure 14 depicts the energy level diagram of the hydrogen atom including all relativistic and spin effects.

The only difference between the results of Sommerfeld and Dirac is that, according to the Dirac theory, the two levels having the same values of n and j but differing in ℓ values are degenerate. This prediction was tested by Lamb and Rutherford[25] who demonstrated that for $n = 2$ and $j = \frac{1}{2}$ there are indeed two levels corresponding to $\ell = 0$ and 1, which have the separation of 4.4×10^{-6} eV with $\ell = 0$ level lying at higher energy relative to the $\ell = 1$ level. This separation is referred to as the Lamb shift, a precise quantitative explanation of which can only be given in terms of the theory of quantum electrodynamics.[35] It appears that the effect of the spin-orbit interaction on the Schrödinger energy levels leads to the correct designation of the split energy levels, although for a precise estimate of the energies one needs to invoke quantum electrodynamics.

The Lamb shift has also been observed in the He^+ ion. The experimental values[36-38] are close to 14,040 MHz, which compare excellently with the theoretical values of 14,042 MHz estimated by Mohr[39] and 14,045 MHz by Erickson.[40]

L. Nuclear Spin and the Magnetic Hyperfine Interaction

Atomic nuclei possess spin angular momentum and spin magnetic moments analogous to those of electrons. The nuclear spin quantum number is denoted by i and can assume integral or half-integral values. For atomic transitions it is sufficient to assume that the nucleus of the atom remains in the ground state and, therefore, i remains constant. Nuclear spin is an intrinsic property of the nucleus and depends on the intrinsic spins of the particles (neutrons and protons) of which a nucleus is composed and on the currents circulating in the nucleus due to the motion of the protons. The magnitude of the nuclear spin momentum vector $\vec{I}$ of hydrogen is analogous to that of the spin of an electron, so that

$$I^2 = i(i+1)\hbar^2 \tag{172}$$

with

$$i = \frac{1}{2} \tag{173}$$

and the z-component of the nuclear spin is

$$I_z = m_i \hbar \tag{174}$$

where

$$m_i = \pm\frac{1}{2} \tag{175}$$

Also, we expect by observing the relation connecting the electron spin and the electron spin angular momentum (see Equation 127), that a proton would have a magnetic moment

$$\begin{aligned} \vec{\mu}_p &= g_p \frac{e}{2M_p} \vec{I} \\ &= g_p \frac{\mu_N}{\hbar} \vec{I} \end{aligned} \tag{176}$$

where M_p = mass of a proton, μ_N is the nuclear magneton given by

$$\mu_N = \frac{e\hbar}{2M_p} = 0.505038 \times 10^{-26} \text{ amp-m}^2 \tag{177}$$

and g_p is proton spin g-factor. Experimentally, the proton magnetic moment has been deduced to have the value,

$$\mu_p = 2.79275\,\mu_N \tag{178}$$

which also defines g_p in conjunction with Equation 176. The z-component of the magnetic moment (from Equation 176) is given by

$$\mu_{p_z} = g_p \frac{\mu_N}{\hbar} I_z$$

$$= g_p \mu_N m_i \quad (179)$$

It was Pauli[41] who suggested in 1924 that the nuclear angular momentum and nuclear magnetic moments may be responsible for certain features of the so-called hyperfine structure (hfs), which results from the interaction of the nuclear magnetic dipole moments with the magnetic field produced by the atomic electrons. The hyperfine interaction energy is expected to be about 2000 times smaller than the ordinary fine-structure splitting as the nuclear magneton is about 2000 times smaller than the Bohr magneton. Owing to this small change in the fine-structure lines the interaction is termed the hyperfine interaction.

The new total angular momentum vector $\vec{F}$ can now be formed by combining the total momentum $\vec{J}$ and the nuclear momentum $\vec{I}$. Accordingly,

$$\vec{F} = \vec{J} + \vec{I} \quad (180a)$$

$$= \vec{L} + \vec{S} + \vec{I} \quad (180b)$$

The magnitude of $\vec{F}$ is given in terms of the associated quantum numbers f by

$$F^2 = f(f+1)\hbar^2 \quad (181)$$

where f takes on integrally spaced values from the smallest positive value $|j - i|$ to the largest positive value $(j + i)$.

In analogy with the spin-orbit interaction, the hyperfine interaction energy may be expressed as

$$(\Delta E)_{hfs} = C\, \vec{J} \cdot \vec{I} \quad (182)$$

where C is a constant proportional to μ_N.

The expectation value of the above interaction can be determined in the same manner as in the case of the spin-orbit interaction treated earlier. Employing the vector combination relation as expressed in Equation 180a, the expectation value of the hyperfine shift is obtained as

$$\langle \Delta E \rangle_{hfs} = C \tfrac{1}{2} [f(f+1) - i(i+1) - j(j+1)] \quad (183)$$

For instance, an energy level corresponding to $j = \frac{1}{2}$ and $i = 3$ would have a hfs composed of two discrete levels characterized by f quantum numbers of 5/2 and 7/2. Similarly a level corresponding to $j = 3/2$ and $i = 3$ is expected to be split by the hyperfine interaction into four lines having f quantum numbers 3/2, 5/2, 7/2, and 9/2.

Hyperfine splitting of atomic energy levels is very difficult to observe experimentally on account of its small magnitude; however it is possible to measure it by optical spectroscopic equipment of extremely high resolution. The magnitude of the hyperfine splitting of hydrogen in its ground state is 1420 megacycles, as determined by modern microwave techniques used in radioastronomy in interstellar space. Such studies are helpful in the investigation of the distribution and dynamics of hydrogen in interstellar and intergalactic space.

The nuclear spin magnetic moment interacts with the external magnetic field, if present, in the same manner as the electron spin magnetic moment interacts with the applied field (see Equation 131). Also, there is the occurrence of the electric hyperfine interaction,[42-44] which is present if the nuclear spin is greater than ½; it is proportional to the electric field gradient produced at the nucleus, which may be due to the external charge distribution. This interaction is written as

$$(\Delta E) = A\,[3m_i^2 - i\,(i+1)]$$

where A is a constant that depends on the quadrupole moment of the nucleus and the electric field gradient at the nuclear site.

M. Parity

A coordinate inversion through the origin changes the coordinates x,y,z to −x,−y,−z (i.e., $\vec{r}$ to $-\vec{r}$). Eigenfunctions which change sign under an inversion of the coordinates are said to be of odd parity, and eigenfunctions which do not change sign under the coordinate inversion are of even parity. Mathematically, if the eigenfunction ψ (x,y,z) satisfies the relation

$$\psi\,(-x,-y,-z) = +\,\psi\,(x,y,z) \tag{184}$$

then the function ψ possesses even parity. Similarly, if

$$\psi\,(-x,-y,-z) = -\psi\,(x,y,z) \tag{185}$$

then ψ has odd parity. All bound or unbound wave functions that are solutions of the time-independent Schrödinger equation for a centrally symmetric potential (such as V(r), in general, and the Coulomb potential, in particular) are of either odd parity or of even parity, i.e., they possess a definite parity. This is a consequence of the fact that the potential at a point $\vec{r}$ is the same as at the point of inversion ($-\vec{r}$), which demands that the probability density $\psi^*\psi$ is the same at $\vec{r}$ and at $-\vec{r}$.

In spherical coordinate systems, the point of inversion of a point r, θ, ϕ, is obtained by the transformation

$$r \to r,\ \theta \to \pi - \theta,\ \phi \to \pi + \phi \tag{186}$$

which is illustrated in Figure 15.

From the properties of the spherical harmonics Y_ℓ^m (θ,ϕ) (see Equation 49) one knows that

$$Y_\ell^m\,(\pi-\theta,\,\pi+\phi) = (-1)^\ell Y_\ell^m\,(\theta,\phi) \tag{187}$$

which follows from the relations

$$P_\ell^m\,[\cos(\pi-\theta)] = (-1)^{\ell-m}\,P_\ell^m(\cos\theta) \tag{188}$$

and

$$e^{im(\pi+\phi)} = (-1)^m\,e^{im\phi} \tag{189}$$

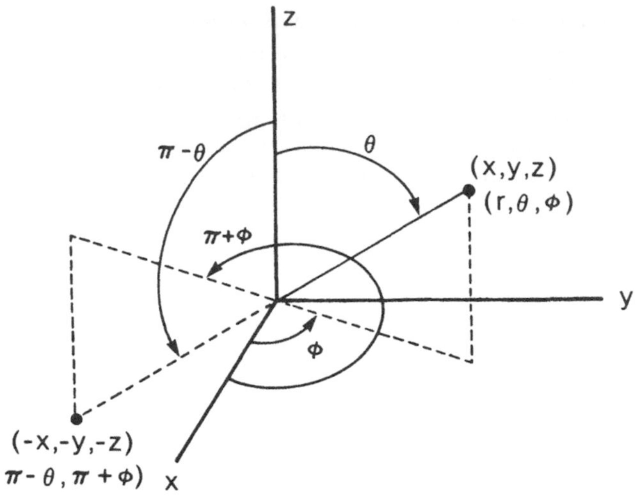

FIGURE 15. Pictorial presentation of the inversion of a point (x, y, z) to obtain (-x, -y, -z) by transformation $r \rightarrow r$, $\theta \rightarrow \pi - \theta$ and $\phi \rightarrow \pi + \phi$ is in a spherical coordinate system.

Hence, the parity of Y_ℓ^m is even or odd according to whether ℓ is even or odd. Since the wave function $\Psi_{n\ell m}$ (Equation 87) is a product of the r-dependent function $R_{n\ell}$ (r) and Y_ℓ^m (θ,ϕ), one also has

$$\Psi_{n\ell m}(r,\pi-\theta,\pi+\phi) = (-1)^\ell \Psi_{n\ell m}(r,\theta,\phi) \tag{190}$$

showing that the parity of $\psi_{n\ell m}$ is odd if ℓ is odd and even if ℓ is even.

The concept of parity plays a fundamental role in studying many topics such as atomic and nuclear radiative transition rates, molecular binding, and excited states. This space-symmetry character of eigenfunctions is also important in understanding the basic interactions among a variety of particles present in nature.

N. Transition Probability and Selection Rules

The process of emission and absorption of radiations by the atomic system constitutes a most characteristic and important phenomenon. One usually treats the problem by first assuming that the atomic system and the radiation field are separate entities and then by introducing the interaction between them as a small perturbation. In general, the interaction depends on the charges, magnetic dipole moments, higher order multipole electric and magnetic moments distributed in the material system, and quantum properties of the electromagnetic radiation field. In the process of interaction the atomic system "jumps" from one energy level to another, emitting or absorbing energy quanta in order to conserve the total energy of the system.

The interaction between a charged particle (electron) and the radiation field can be described by the interaction potential in the first order,

$$V_{int} = \frac{e}{m} \vec{p} \cdot \vec{A} \tag{191}$$

where $(-e)$ is the electron charge, m is the mass of the electron, $\vec{p}$ is the linear momentum of the electron, and $\vec{A}$ is the magnetic vector potential of the radiation at the location of the particle.

The interaction potential in Equation 191 can be derived from the Hamiltonian of a charged particle in the radiation field, which can be obtained from the Hamiltonian of the particle in the absence of the radiation field on replacing the linear momentum operator $\vec{p}$ by $(\vec{p} + (e/c)\vec{A})$. One then retains only the first-order term. The second-order term, which is proportional to A^2, can be neglected since the lowest-order term is dominant. The second-order term is, however, important when such processes as the scattering of light and Raman effect[45] are under consideration.

Because of the interaction with the radiation field, a system which is in the initial quantum state designated by the symbol m has a definite probability of transition to the final quantum state k. The first-order transition probability per unit time W_{km} is given by the Fermi golden rule[15,46]

$$W_{km} = \frac{2\pi}{\hbar} \rho(k) |\langle k | V_{int} | m \rangle|^2 \tag{192}$$

where $\varrho(k)$ is the energy density of the final states. The symbol $\langle k|V_{int}|m\rangle$ is termed as a matrix element of V_{int} between the initial state ψ_m and the final state ψ_k, and represents the integral

$$\langle k | V_{int} | m \rangle = \int \Psi_k^* V_{int} \Psi_m \, d\tau$$

where $d\tau$ is the volume element.

In the present case the system is comprised of an atomic system and the radiation field interacting by the potential of Equation 191. If a photon is emitted by the atom under the influence of the applied radiation field, the relevant phenomenon is known as stimulated emission. On the other hand, the phenomenon of emission in the absence of the applied radiation field is called spontaneous emission.

Quantum electrodynamics can explain the phenomenon of spontaneous emission observed in nature in which there always exists electromagnetic radiations, even in the absence of an applied field. Since the electromagnetic field has discretely quantized energy levels, including the zero-point energy level, according to quantum electrodynamics the required frequency is always present in the space surrounding the atomic system to induce charge oscillations in the system which, in turn, cause the atom to radiate spontaneously.

Considering the normal frequencies of the radiation field and the quantum states of the atomic system, one simplifies Equation 192 by what is known as the dipole approximation to obtain the transition rate for the transition of the atomic state a to the state a′,

$$W_{a'a} = \frac{2}{\pi\epsilon_0 \hbar c^3} \omega^3 |\langle a' | P_s | a \rangle|^2 (\bar{n}+1) \tag{193a}$$

for the emission of a photon, and

$$W_{a',a} = \frac{2\,\omega^3}{\pi\epsilon_0 \hbar c^3} |\langle a' | P_s | a \rangle|^2 \, \bar{n} \tag{193b}$$

for the absorption of a photon.

In Equations 193, $\bar{n}$ is the average excitation of the normal modes of the radiation field in the neighborhood of the frequency ω which is

$$\omega = (E_{a'} - E_a)/\hbar \qquad (194)$$

E_a and $E_{a'}$ being the energies of the atomic system in the initial and the final state. P_s is the component of the electric dipole moment $\vec{P}$ of the atomic system along the direction of the electric field of the emitted or absorbed photon. In the case of a number of charged particles the dipole moment $\vec{P}$ is the sum of the individual dipole moments $\vec{P}_i$ such that

$$\vec{P} = \sum_i \vec{P}_i = \sum_i e_i \vec{r}_i \qquad (195)$$

where e_i is the charge of the particle i located at the position $\vec{r}_i$. For spontaneous emission, $\bar{n} = 0$ and then the average power radiated by the system is

$$R = \hbar\omega W_{a',a}$$

$$= \frac{2}{3} \frac{\omega^4}{\pi \epsilon_o c^3} \sum_{s=1}^{3} |<a'|P_s|a>|^2 \qquad (196)$$

In Equation 196 the last step makes use of the average of $|<a'|P_s|a>|^2$ over all directions of the polarization of the emitted photon. Equation 196 can be compared with the power emitted by a classical oscillating electric dipole $\vec{P} = \vec{P}_o \cos \omega t$, which is

$$R_{c\ell} = \frac{1}{3} \frac{\omega^4}{\pi \epsilon_o c^3} P_o^2 \qquad (197)$$

On comparison of Equation 196 with 197, one finds that a quantum mechanical system emits radiations at the same rate as a classical oscillating electric dipole of strength P_o such that

$$P_o^2 = 2 \sum_{s=1}^{3} |<a'|P_s|a>|^2 \qquad (198a)$$

$$= 2e^2 [|<a'|x|a>|^2 + |<a'|y|a>|^2 + |<a'|z|a>|^2] \qquad (198b)$$

P_o is known as the oscillator strength corresponding to the atomic transition $a \rightarrow a'$.

It is obvious from Equation 193a that the transition rate for the stimulated emission of the photon is proportional to the intensity of the applied radiation field. For fields of high intensity the transition rate is extremely high and the atoms radiate photons with large efficiency. Indeed, lasers generate extremely bright beams of coherent light in this manner.

Equation 196 is very useful in revealing important features of the spectroscopic transitions. The intensity of the emitted light is proportional to the absolute square of the dipole transition matrix $<a'|P_s|a>$, which is different for different possible transitions. This explains the wide variation in intensity exhibited by different spectral lines associated with these transitions. Sometimes it happens that the matrix elements $<a'|P_s|a>$ vanish for some values of a and a'. In that case the transition from the state a to a' is not possible (which is strictly true in the approximation of the dipole transition represented by Equation 196). This gives rise to what is known as the selection rules for the transition between the states a and a′. A transition from the state a to a′ in which the matrix element $\langle a'|P_s|a\rangle$ vanishes is known as a first-order forbidden transition.

Also, Equation 194 embodies the essential condition that energy be conserved in the transition. This inference is true for first-order transitions of the type we have considered. In the higher order approximation it could be that a direct transition from the lower state a to the excited state a′ is not possible, but when sufficient energy is available the system could jump by direct transition from the state a to the higher excited state b (if it exists) from which it could reach the lower state a′ — again by a direct transition. In this situation, instead of the matrix element $\langle a'|P_s|a\rangle$ for the direct single transition in the first-order, one would have the product of the matrix elements $\langle a'|P_s|b\rangle \langle b|P_s|a\rangle$ in the second-order, leading to the double transition. Over a long time in the double transition, energy is conserved by emitting a photon of energy $E_{a'} - E_a$, but because of the involvement of the state b it need not be that the energy is conserved at all times. Processes of this sort, where it appears that the system has violated energy conservation temporarily, are referred to as virtual processes. Schematic illustrations of the virtual and direct processes discussed above have been presented in Figure 16.

Next we turn to the deduction of the electric dipole selection rules[47] for one-electron atoms. The selection rules are the set of conditions imposed on the quantum numbers of the initial and final states for observing the nonvanishing transition.

For finding the probability of transition of an atom from the initial state $\psi_{n\ell m}$ to the state $\psi_{n'\ell'm'}$ we must evaluate the matrix elements appearing in Equations 198. The first matrix element is

$$X = \langle n'\ell'm'_{\ell'}|x|n\ell m_\ell\rangle$$

$$= A^*_{n'\ell'm'_{\ell'}} A_{n\ell m_\ell} \int_0^\infty \int_0^\pi \int_0^{2\pi}$$

$$R_{n'\ell'}(r)\, Y_{\ell'}^{m'_{\ell'}*}(\theta,\phi)\ [r \sin\theta \cos\phi]$$

$$R_{n\ell}(r)\, Y_\ell^{m_\ell}(\theta,\phi)\, r^2 dr\ \sin\theta\, d\theta\, d\phi \qquad (199)$$

By direct evaluation[48] of the above expression or by parity considerations of the spherical harmonics $Y_{\ell'}^{m_{\ell'}}$ and $Y_\ell^{m_\ell}$, one obtains that

$$X \neq 0$$

if

$$m_\ell - m_{\ell'} = \Delta m_\ell = \pm 1 \qquad (200a)$$

and

$$\ell - \ell' = \Delta\ell = \pm 1 \qquad (200b)$$

Similarly,

$$Y = \langle n'\ell'm'_{\ell'}|y|n\ell m_\ell\rangle \neq 0$$

if

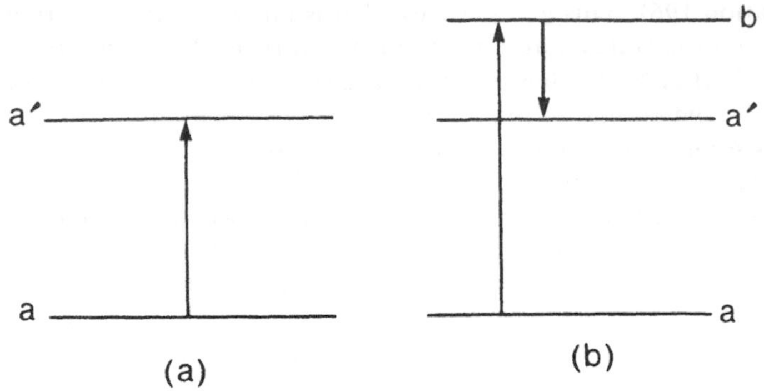

FIGURE 16. Schematic illustration of the transition in (a) a direct process and in (b) a virtual process from the state a to a′. The direct transition from a to a′ is possible when $<a'|P_s|a> \neq O$. In a virtual transition the matrix $<a'|P_s|a> = O$, and if a higher state b is present the system makes the transition from the state a to the state b and then from the state b to the state a′.

$$m_\ell - m_\ell' = \Delta m_\ell = \pm 1 \qquad (201a)$$

and

$$\ell - \ell' = \Delta\ell = \pm 1 \qquad (201b)$$

and

$$z = < n'\ell'm'_{\ell'} | y | n\ell m_\ell > \neq 0$$

if

$$m_\ell - m_{\ell'} = \Delta m_\ell = 0 \qquad (202a)$$

and

$$\ell - \ell' = \Delta\ell = \pm 1 \qquad (202b)$$

Summarizing the above results we obtain the electric dipole selection rules

$$\Delta\ell = \pm 1 \qquad (203a)$$

and

$$\Delta m = 0 \text{ or } \pm 1 \qquad (203b)$$

which are the conditions for which the transitions can occur. If one evaluates the dipole matrix elements from the known wave functions it is possible to obtain the relative intensities of various spectral lines of hydrogen and other one-electron atoms by using Equation 196.

The above selection rules were derived by omitting the presence of the spin angular momentum of the electron. If one considers the spin quantum number s of the electron, one obtains the selection rules:

$$\Delta \ell = \pm 1$$

$$\Delta s = 0$$

$$\Delta j = \pm 1 \text{ or } 0 \text{ but } j = 0 \nleftrightarrow j' = 0$$

$$\Delta m_j = \pm 1 \text{ or } 0 \text{ but } m_j = 0 \nleftrightarrow m_{j'} = 0 \text{ if } \Delta j = 0 \qquad (204)$$

In the above we have considered only the electric dipole transitions. Radiative transitions which are analogous to the classical oscillating magnetic dipoles, electric or magnetic quadrupoles, or other higher order multipoles are also possible. Sometimes these transitions may be responsible for the intense lines in an atomic spectrum. The higher order multipole transitions are very important for the radiative transitions in the case of nuclei.

The transition rate R (see Equation 196) for the hydrogen atom for a transition from its first excited state prescribed by the principal quantum number n = 2 to the ground state (n = 1), comes out to be R ≅ 10^8 sec^{-1}. This means that in about 10^{-8} sec one transition may occur. It is because of this that the first excited state is said to have the lifetime t = 10^{-8} sec. In the situation illustrated in Figure 16b the transition from state a′ to a is inhibited, with the result that the system remains in the excited state a′ for an appreciable fraction of a second compared with the normal lifetime 10^{-8} sec. The excited state is then referred to as a metastable state. The delayed emission of light quanta from the metastable state gives rise to the phenomenon known as phosphorescence. In the case of certain atomic species and in the case of nuclear transitions, it has been possible to observe this phenomenon.

O. Comparison of the Old and Modern Quantum Theories

It is interesting to compare the results of the old theory of Bohr and Sommerfeld and the modern quantum theory developed by Schrödinger, Dirac, and others, which also include quantum electrodynamics. The old theory of Bohr is capable of explaining the unsplit hydrogen atom energy levels. The relativistic modification by Sommerfeld gives the correct fine-structure splitting of hydrogen, though it is found to be accidental. In terms of the old theory it was not possible to account for hyperfine splitting, the Lamb shift of the hydrogen atom, and the selection rules and transition rates for the emission and absorption of radiations. Also, the old theory failed completely if applied to multielectron atoms.

On the other hand, the modern theory has been able to explain from the broad features to the fine details the atomic spectra of one-electron and multielectron atoms. It has accounted for the fine structure, hyperfine structure, and Lamb shift of the hydrogen atom. In fact, a highly refined calculation for a system can be made by including more and more higher order interactions in the framework of the Schrödinger theory. In particular, the modern quantum theory has been successful in interpreting the energy levels and fine and hyperfine structure of multielectron atoms. It lays down the basic theory for explaining the selection rules and the transition rates for the spectra emitted by all the one-electron and multielectron atoms. As we shall see in the following, the Schrödinger theory can be applied to the multielectron atom in the same fashion as in the case of the single-electron atom. We shall find that the theory leads to the proper explanation of detailed atomic structure, though the calculations involved are more complicated than for a single-electron atom.

III. QUANTUM MECHANICS OF MULTIELECTRON ATOMS

A. Introduction

In this section we shall be concerned with the application of the Schrödinger equation to atoms which contain more than one electron. We are required to deal with a system of identical particles (which are electrons in the present case) that leads us to what is known as the Pauli exclusion principle. The phenomenon of exchange forces which does not appear in the classical treatment will be discussed here. The detailed treatment of the helium atom, a system of two electrons bound to a nucleus of positive charge 2e, will be given as an example to illustrate the basic principles involved in the multielectron system. The ground state of the multielectron system and the shell model of the atomic system will be considered; these will be helpful in understanding the periodic nature of the elements. Thus, the quantum mechanical basis for the periodic table will be made explicit. This will also lay down the background for understanding important fields such as organic and inorganic chemistry, molecular physics, solid state physics, molecular biology, etc. The high-energy excited states of multielectron atoms for the production of X-rays will be treated in Section IV.

Important theoretical developments will also be presented for multielectron systems. We shall find that it is possible to obtain quantitative information about atoms from the calculations performed on high-speed electronic computers. Indeed, the explanation of various observed results by calculations for atoms provide the most satisfying triumphs of modern quantum theory.

B. Identical Particles, Exchange Symmetry of Wave Functions and the Exclusion Principle

Electrons form a system of identical particles; any one electron is exactly like any other electron. Quantum mechanics, unlike the classical mechanics, requires special treatment for identical particles because the Heisenberg uncertainty principle plays an important role in such a system. In the classical situation it is possible to follow exactly the motion of the individual particles by labeling all the identical particles. However, the situation is different in a quantum phenomenon involving identical particles since, owing to the uncertainty principle, the behavior of the particles is disturbed in the process of measurement, consequently making the particles indistinguishable. For instance, if photons are used to observe the electrons, the photons interact with the electrons and disturb their motion severely in an unpredictable way causing the situation in which the electrons cannot be identified exactly. A quantum mechanical description can be developed by considering that the state of a particle is given by its wave function and, in the case of more than one identical particle, the wave functions associated with the particles overlap with one another, thereby making it impossible to determine which wave function is exactly associated with a particular particle. This also restricts the mathematical form of the wave functions for the quantum mechanical description of identical particles, as will be apparent shortly.

For a system of N identical particles, the Hamiltonian can be written as

$$\mathcal{H} = -\frac{\hbar^2}{2m}(\nabla_1^2 + \nabla_2^2 + \ldots \nabla_N^2) + V(\vec{r}_1, \vec{r}_2 \ldots \vec{r}_N) \qquad (205)$$

where $(-\hbar^2/2m)\nabla_i^2$ expresses the kinetic energy operator of the ith particle having the coordinates, $\vec{r}_i = (x_i, y_i, z_i)$. If particles 1 and 2 are interchanged, the Hamiltonian becomes

$$\mathcal{H}' = -\frac{\hbar^2}{2m}(\nabla_2^2 + \nabla_1^2 + \ldots \nabla_N^2) + V(\vec{r}_2, \vec{r}_1, \ldots, \vec{r}_N) \quad (206)$$

Because the particles are identical, the potential energy does not change and hence

$$\mathcal{H}' = \mathcal{H} \quad (207)$$

Let us now define an operator P_{rs}, called the exchange operator, whose action is to exchange a pair of particles r and s if it operates on any function of the coordinates of the particles. Accordingly,

$$P_{rs}\,\psi(1, 2, \ldots, r, \ldots, s, \ldots, N) = \psi(1, 2, \ldots, s, \ldots, r, \ldots, N) \quad (208)$$

where we have introduced a simplified notation in which the coordinates (including spin) of the ith particle are denoted by i. It is clear that the double application of P_{rs} on the wave function leaves the wave function unchanged. That is,

$$P_{rs}^2\,\psi = \psi_{rs} \quad (209)$$

which shows that the eigenvalues of the operator P_{rs} are ± 1,

$$P_{rs} = \pm 1 \quad (210)$$

Also, applying the exchange operator to both sides of the Schrödinger equation,

$$\mathcal{H}\,\psi = E\,\psi \quad (211)$$

and employing Equation 207 (or directly from Equation 210), one obtains

$$P_{rs}\,\psi = \pm\,\psi \quad (212)$$

or

$$\psi(1, 2, \ldots r, \ldots s, \ldots N) = \pm\psi(1, 2, \ldots, s, \ldots, r, \ldots N) \quad (213)$$

Thus the eigenfunctions ψ are either symmetric or antisymmetric under the exchange of any two particles. It must be stressed that the exchange of particles incorporates the exchange of space coordinates as well as spins. The exchange of space and spin coordinates has been depicted in Figure 17.

It has been found experimentally that identical particles which possess half-integral quantum numbers for their intrinsic spins are to be described by the wave functions which are antisymmetric under the exchange of any two particles. Such particles are known as Fermi particles since they follow Fermi-Dirac statistics.[15] For example, electrons, nucleons (protons and neutrons), and μ-mesons are spin ½ particles and follow Fermi-Dirac statistics, and therefore the condition pertains:

$$\psi(1, 2, \ldots, r, \ldots, s, \ldots N) = -\psi(1, 2, \ldots, s, \ldots, r, \ldots N) \quad (214)$$

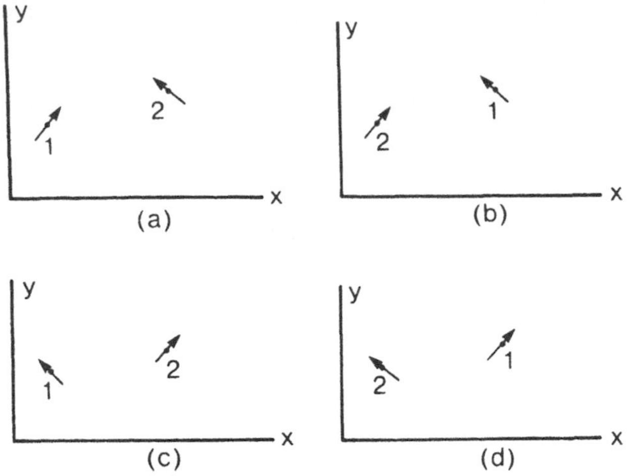

FIGURE 17. Exchange of space and spin coordinates of two particles, 1 and 2, with spins as shown by arrows: (a) initial configuration, (b) configuration when space coordinates are exchanged, (c) configuration when spin coordinates are exchanged, and (d) configuration when both space and spin coordinates are exchanged.

On the other hand, identical particles that have integral spin quantum numbers are described by wave functions which are symmetric under the exchange of any two particles. Such particles are termed Bose particles since they follow Bose-Einstein statistics. For example, photons have spin 1 and neutral helium atoms in their ground state and α particles have spin 0; therefore they are Bose particles whose wave functions satisfy the following symmetric property under the exchange of any two particles:

$$\psi(1, 2, \ldots, r, \ldots, s, \ldots N) = \psi(1, 2, \ldots, s, \ldots, r, \ldots N) \quad (215)$$

Such symmetry restrictions on the wave functions have very important physical consequences. The most important one is that Bose particles exhibit an additional attraction between the particles and tend to cluster near one another in space. On the contrary, Fermi particles repel one another and have the tendency not to occupy the same position. Also, because of the symmetric and antisymmetric nature of the wave functions under the exchange of particles, Bose particles tend to occupy the same quantum states whereas Fermi particles avoid occupying states with the same quantum numbers. The latter deduction is known as Pauli's exclusion principle.[41] Pauli postulated this principle to explain the electronic structure of atoms; it states that no two electrons in an atom can have the same set of quantum numbers n,ℓ,m_ℓ,m_s. As will be seen later, Pauli's exclusion principle is of utmost importance for the periodic system of elements. Indeed, if electrons did not follow the Pauli principle we would have an entirely different universe; atoms would all be inert, thus molecules would not form, and without molecules there would be no life.

It must be remarked that the Pauli exclusion principle is a special case of the general requirement that the wave function of a system of electrons is antisymmetric with the exchange of any two electrons. This stronger antisymmetric property should be satisfied in accurate quantum mechanical calculations, though in approximate calculations the Pauli principle can be satisfactory.

An energy eigenfunction for a system of fermions (Fermi-particles) containing N particles can be written in the form of a determinant known as the Slater determinant,

$$\psi(1, 2, \ldots N) = A \begin{vmatrix} \psi_a(1) & \psi_a(2) & \psi_a(N) \\ \psi_b(1) & \psi_b(2) & \psi_b(N) \\ \cdot & \cdot & \cdot \\ \cdot & \cdot & \cdot \\ \cdot & \cdot & \cdot \\ \cdot & \cdot & \cdot \\ \psi_n(1) & \psi_n(2) & \psi_n(N) \end{vmatrix} \quad (216)$$

where $\psi_a(1)$ describes a wave function which contains particle 1 in state a, $\psi_b(2)$ describes particle 2 in state b, and so on. The constant A is determined by the normalization condition

$$\int \psi^*(1, 2, \ldots N)\, \psi(1, 2, \ldots N)\, d^3r_1\, d^3r_2 \ldots d^3r_N = 1 \quad (217)$$

which is usually abbreviated in the Dirac notation,

$$< \psi | \psi > = 1 \quad (218)$$

If the one-electron functions $\psi_j(i)$ in Equation 216 are orthonormal, that is, they satisfy the relation

$$< \psi_j(i) | \psi_{j'}(i) > = \int \psi_j^*(i)\, \psi_{j'}(i)\, d^3r_i = \delta_{j,j'} \quad (219)$$

(δ, in general,$_{i,j}$ is the Krönecker δ-function which is 0 if $i \neq j$ and 1 if $i = j$) then Equation 217 yields

$$|A| = \frac{1}{\sqrt{N!}} \quad (220)$$

The Slater determinant (Equation 216) satisfies the antisymmetry postulate for fermions and identically vanishes if two particles occupy the same state.

Let us now turn to a system of two electrons. We assume for the moment that they do not interact with one another by any interaction force (such as the Coulomb force) and that their spins are not coupled by any external field. In this situation the Hamiltonian operator is invariant with respect to (1) the exchange of all the coordinates, space as well as spin, of the two electrons, (2) exchange of space coordinates alone of the two electrons, and (3) exchange of spin coordinates alone. These symmetry properties of the Hamiltonian lead one to write the total wave function of the system as a product of space-dependent and spin-dependent functions, such that

$$\Psi = \psi(\vec{r}_i)\, \chi\, (s_i) \quad (221)$$

where Ψ is the total wave function, $\psi(\vec{r}_i)$ depends on the space coordinates $\vec{r}_i$ alone, and $\chi(s_i)$ on the spin coordinates s_i alone. Because of the symmetry properties of the Hamiltonian operator mentioned above, the functions Ψ and χ should be of the form

$$\psi^{\pm}(\vec{r}_i) = \frac{1}{\sqrt{2}} [\psi_a(\vec{r}_1)\psi_b(\vec{r}_2) \pm \psi_a(\vec{r}_2)\psi_b(\vec{r}_1)] \tag{222}$$

and

$$\chi^{\mp}(s_i) = \frac{1}{\sqrt{2}} [\chi_a(s_1)\chi_b(s_2) \mp \chi_b(s_1)\chi_a(s_2)] \tag{223}$$

The factors $\frac{1}{\sqrt{2}}$ appear because of the normalization of Ψ and χ functions. The product function Ψ in Equation 221 is then formed by the requirement that the total wave function for fermions (bosons) must be antisymmetric (symmetric) in the exchange of all coordinates — space and spin — of the two electrons; so that the signs in Equations 222 and 223 are the opposite (same) in forming the product function Ψ.

C. Exchange Forces, Exchange Degeneracy, and Exchange Energy

For the case of two noninteracting particles the total wave function is given by Equations 221, 222, and 223. To deduce an important physical effect let us evaluate the expectation value of the square of the distance between the two particles. Making use of Equation 221 to 223 we have

$$\begin{aligned} \langle \Psi | (\vec{r}_2 - \vec{r}_1)^2 | \Psi \rangle &= \langle \psi_a | r^2 | \psi_a \rangle + \langle \psi_b | r^2 | \psi_b \rangle \\ &\quad - 2 \langle \psi_a | \vec{r} | \psi_a \rangle \cdot \langle \psi_b | \vec{r} | \psi_b \rangle \\ &\quad \mp 2 | \langle \psi_a | \vec{r} | \psi_b \rangle |^2 \end{aligned} \tag{224}$$

where the negative (positive) sign in the last term corresponds to the symmetric (antisymmetric) wave function in Equation 222. Analysis of Equation 224 reveals that the first three terms are present even when the two particles are distinguishable, that is, when their wave function is simply $\psi_a(1)\psi_b(2)$. The fourth term appears because of the indistinguishability of the two particles. It is obvious from Equation 224 that when the wave function is symmetric the distance between the two particles is reduced, and when it is antisymmetric the distance is increased. It means that in the symmetric case the particles attract one another whereas in the antisymmetric case they repel. These attraction and repulsion effects are due to what is known as exchange forces and their origin is purely quantum mechanical; they do not have a classical analogue and arise from the characteristics of the spin-angular momenta. The exchange forces are physically so real that it has been shown by Heisenberg[32,49,50] that one could very well introduce in the Hamiltonian function a term of the simplified form

$$\mathcal{H}_{ex} = \pm J_{12} \vec{s}_1 \cdot \vec{s}_2 \tag{225}$$

where J_{12} is the exchange coupling constant. It must be emphasized that the exchange forces are always present between identical particles, even when the particles are noninteracting. Explicitly, two electrons, even if thought to have no Coulomb repulsion, are subjected to the quantum mechanical exchange forces. Such forces also arise between two neutrons or between two protons.

For an electron the spin quantum numbers are $s = \frac{1}{2}$ and $m_s = \pm \frac{1}{2}$. We shall employ a common abbreviated notation for these two states, viz., the α and β symbols which have the following correspondence,

$$|s{=}\tfrac{1}{2},\ m_s{=}\ \tfrac{1}{2}> \ = \alpha \qquad (226a)$$

$$|s{=}\tfrac{1}{2},\ m_s = -\tfrac{1}{2}> \ = \beta \qquad (226b)$$

The spin functions α and β are frequently referred to as spin "up" and spin "down" states, respectively. Using the notations such as $\alpha(1)$, which means that the particle 1 is assigned to state α, and so on, one constructs the symmetric and antisymmetric spin states for a system of two particles as

$$\chi^+ = \begin{cases} \alpha(1)\,\alpha(2) \\ \dfrac{1}{\sqrt{2}}\,[\alpha(1)\,\beta(2) + \alpha(2)\,\beta(1)] \\ \beta(1)\,\beta(2) \end{cases} \qquad (227)$$

and

$$\chi^- = \frac{1}{\sqrt{2}}\,[\alpha(1)\,\beta(2) \ - \ \alpha(2)\,\beta(1)] \qquad (228)$$

where χ^+ is the spin-symmetric and χ^- spin-antisymmetric wave functions. Thus for a two-electron system there are four antisymmetric wave functions (see Equations 221, 222, 227, and 228),

$$\Psi_T = \frac{1}{\sqrt{2}}\,[\psi_a(1)\Psi_b(2) - \Psi_a(2)\Psi_b(1)]$$

$$\begin{cases} \alpha(1)\,\alpha(2) \\ \dfrac{1}{\sqrt{2}}\,[\alpha(1)\,\beta(2) + \alpha(2)\,\beta(1)] \\ \beta(1)\,\beta(2) \end{cases} \qquad (229)$$

$$\Psi_S = \frac{1}{\sqrt{2}}\,[\psi_a(1)\,\psi_b(2) + \psi_a(2)\,\psi_b(1)]$$

$$\frac{1}{\sqrt{2}}\,[\alpha(1)\,\beta(2) - \alpha(2)\,\beta(1)] \qquad (230)$$

where Ψ_T is known as the triplet state and Ψ_S the singlet state. The total spin angular momentum of the system is

$$\vec{S} = \vec{S}_1 + \vec{S}_2$$

It can be verified directly by using the spin functions of Equations 227 and 228 that χ^+ corresponds to the total angular momentum quantum numbers S = 1 and M_s = 1,0,−1 whereas χ^- corresponds to S = 0 and M_s = 0.

If the two electrons are not interacting through any force, the four functions expressed in Equations 229 and 230 correspond to the same eigenvalue. This is known as exchange degeneracy. Let us now add the Coulomb interaction term $e^2/4\pi\varepsilon_0 r_{12}$ between the electrons in the Hamiltonian. Though this term does not depend explicitly on the spin coordinates of the electrons, the energy eigenvalues corresponding to the four functions given in Equations 229 and 230 are no longer the same. This will partially remove the energy degeneracy. Explicitly, since the electrons are closer, on the average, when they are in the singlet state, the Coulomb interaction increases the energy of the singlet state more than that of the triplet state (see Figure 18). This gives rise to the exchange splitting between the singlet and triplet states, and the energy difference between the singlet and triplet states is known as the exchange energy. Though the exchange energy is an electrostatic energy, it appears because of the exchange force which causes the aligned spin state to attain lower energy. Consequently, a parallel alignment of spins is energetically more favorable for electrons in the same orbital state. Thus exchange forces tend to align the spins of those electrons which occupy the same orbital without violating the Pauli exclusion principle. This is the physical origin of what is known in atomic physics as Hund's rule.[51,52]

D. Helium Atom

A helium atom consists of two electrons and a relatively massive nucleus of positive charge Ze with Z = 2. Assuming that the nucleus is stationary, the Hamiltonian for the helium atom is

$$\mathcal{H} = \frac{p_1^2}{2m} - \frac{e^2}{4\pi\epsilon_0 r_1} + \frac{p_2^2}{2m} - \frac{e^2}{4\pi\epsilon_0 r_2} + \frac{e^2}{4\pi\epsilon_0 |\vec{r}_1 - \vec{r}_2|} \qquad (231)$$

where $p_1^2/2m$ and $p_2^2/2m$ are the kinetic energies of the electrons, and $\vec{r}_1$ and $\vec{r}_2$ are the radius vectors of the electrons measured from the nucleus. If the last term, the Coulomb repulsion between the electrons is neglected, the zero-order Hamiltonian is

$$\mathcal{H}_0 = \frac{p_1^2}{2m} - \frac{e^2}{4\pi\epsilon_0 r_1} + \frac{p_2^2}{2m} - \frac{e^2}{4\pi\epsilon_0 r_2} \qquad (232)$$

which is the sum of two hydrogen-atom-type Hamiltonians. The eigenfunctions and eigenvalues of $\mathcal{H}_0$ are

$$\psi^{(0)} = \psi_{n\ell m}(\vec{r}_1)\,\psi_{n'\ell'm'}(\vec{r}_2) \qquad (233)$$

and

$$E^{(0)} = -\frac{mZ^2e^4}{(4\pi\epsilon_0)^2 2\hbar^2}\left(\frac{1}{n^2} + \frac{1}{n'^2}\right)$$

$$= -Z^2E_0\left(\frac{1}{n^2} + \frac{1}{n'^2}\right) \qquad (234a)$$

where

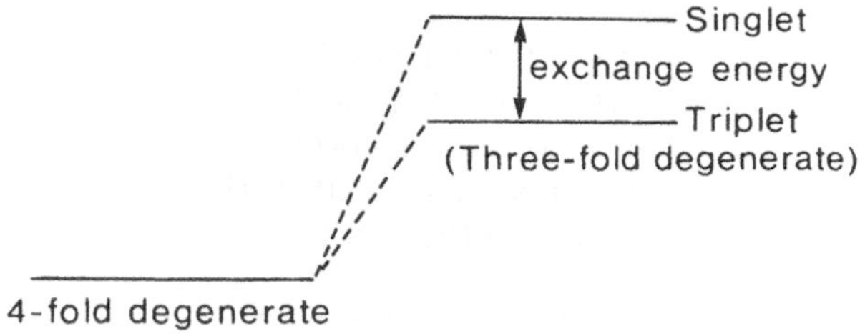

FIGURE 18. Exchange splitting of the energy of a two-electron system due to the Coulomb repulsion.

$$E_O = me^4/2\hbar^2(4\pi\epsilon_O)^2 \tag{234b}$$

For a helium atom, the ground state corresponds to n = 1 and n′ = 1. The lowest energy configuration is obtained when both electrons occupy the lowest energy spatial wave functions $\psi_{100}(r)$. Then the symmetric and antisymmetric two-particle spatial wave functions are (see Equation 222)

$$\psi^+_{100,100}(\vec{r}_1, \vec{r}_2) = \sqrt{1/2}\,[\psi_{100}(\vec{r}_1)\psi_{100}(\vec{r}_2) + \psi_{100}(\vec{r}_2)\psi_{100}(\vec{r}_1)] \tag{235}$$

and

$$\psi^-_{100,100}(\vec{r}_1, \vec{r}_2) = \sqrt{1/2}\,[\psi_{100}(\vec{r}_1)\psi_{100}(\vec{r}_2) - \psi_{100}(\vec{r}_2)\psi_{100}(\vec{r}_1)] \tag{236}$$

The antisymmetric space function ψ^- vanishes when both of the electrons occupy the same space function. Therefore, the ground state of the helium atom corresponds to ψ^+, the symmetric space state. To satisfy the requirement that the total wave function must be antisymmetric, one writes the total ground state wave function by multiplying the space wave function with the antisymmetric spin wave function. Accordingly, the ground state of the helium atom is the singlet state given by

$$\psi_O(\vec{r}_1 s_1, \vec{r}_2 s_2) = \psi^+_{100,100}(\vec{r}_1, \vec{r}_2)\frac{1}{\sqrt{2}}[\alpha(1)\,\beta(2) - \alpha(2)\,\beta(1)] \tag{237}$$

The energy of the lowest state is then

$$E_{100,100} = -Z^2E_O - Z^2E_O + \left\langle \psi^+_{100,100}(\vec{r}_1\vec{r}_2) \left| \frac{e^2}{4\pi\epsilon_O}\frac{1}{|\vec{r}_1 - \vec{r}_2|} \right| \psi^+_{100,100}(\vec{r}_1\vec{r}_2) \right\rangle \tag{238}$$

where the spin wave functions integrate out to unity. The computation of the last term in Equation 238 yields

Table 4
COMPILATION OF THEORETICAL AND EXPERIMENTAL GROUND STATE ENERGIES OF THE He-TYPE IONS

		$E_{100,100}$ (eV)	
Z	System	Theory	Experiment
3	Li^{+}	−192.80	−197.14
4	Be^{2+}	−365.31	−369.96
5	B^{3+}	−591.94	−596.4
6	C^{4+}	−872.69	−876.2

$$\left\langle \Psi^{+}_{100,100}(\vec{r}_1 \vec{r}_2) \left| \frac{e^2}{4\pi\epsilon_0} \frac{1}{|\vec{r}_1 - \vec{r}_2|} \right| \psi^{+}_{100,100}(\vec{r}_1, \vec{r}_2) \right\rangle$$

$$= \frac{5}{4} ZE_0 \quad (239)$$

Thus, the ground state energy of the helium atom becomes

$$E_{100,100} = -2Z^2E_0 + \frac{5}{4} ZE_0 \quad (240)$$

which, for Z = 2, gives −74.42 eV as compared with the experimental value −78.62 eV, a remarkably good agreement. Other two-electron systems are singly ionized Li, doubly ionized Be, triply ionized B, etc. The calculated values for these systems from Equation 240 have been compared with the corresponding experimental values in Table 4 and show strikingly good results from the above treatment.

The ionization energy of the He atom is given by the difference between the ground state energy as given in Equation 240 and the energy of the corresponding one-electron ion and a free electron without any kinetic energy. Thus

Ionization energy of He atom

$$= -Z^2E_0 - \left[-2Z^2E_0 + \frac{5}{4} ZE_0\right]$$

$$= Z^2E_0 - \frac{5}{4} ZE_0 \quad (241)$$

One can also calculate the excited states of the helium atom following the above treatment. The excited state can be obtained by assuming that one electron is in the state ψ_{100} and the other electron is in some general excited state $\psi_{n\ell m}$. In this circumstance, the total antisymmetric wave functions can be constructed as before (see Equations 229 and 230) to obtain

$$\psi_S = \psi^{+}_{100,n\ell m}(\vec{r}_1, \vec{r}_2)\ \chi^{-}(s_1, s_2) \quad (242)$$

$$\psi_T = \psi^{-}_{100,n\ell m}(\vec{r}_1, \vec{r}_2)\ \chi^{+}(s_1, s_2) \quad (243)$$

where $\psi_{\pm 100,n\ell m}(\vec{r}_1,\vec{r}_2)$ are formed from the one-electron hydrogen-like functions ψ_{100} and $\psi_{n\ell m}$ as in Equations 235 and 236. The expectation values of the Hamiltonian (Equation 231) can now be calculated separately for the triplet and singlet wave functions ψ_T and ψ_s given in Equations 242 and 243. One obtains

$$E_{100,n\ell m} = -Z^2E_o - \frac{Z^2E_o}{n^2} + J_{100,n\ell m} \pm K_{100,n\ell m} \tag{244}$$

where the positive sign corresponds to the singlet state and the negative sign to the triplet state. J and K are the Coulomb and exchange integrals, respectively, defined by,

$$J_{100,n\ell m} = \iint \psi^*_{100}(\vec{r}_1)\psi^*_{n\ell m}(\vec{r}_2) \frac{e^2}{4\pi\epsilon_o|\vec{r}_1-\vec{r}_2|} \psi_{100}(\vec{r}_1)\psi_{n\ell m}(\vec{r}_2)\, d^3r_1 d^3r_2 \tag{245}$$

and

$$K_{100,n\ell m} = \iint \psi^*_{100}(\vec{r}_1)\psi^*_{n\ell m}(\vec{r}_2) \frac{e^2}{4\pi\epsilon_o|\vec{r}_1-\vec{r}_2|} \psi_{100}(\vec{r}_2)\psi_{n\ell m}(\vec{r}_1)\, d^3r_1 d^3r_2 \tag{246}$$

The integrals J and K are found to be positive, which is clear from the fact that they represent the expectation value of $e^2/4\pi\varepsilon_o|\vec{r}_1 - \vec{r}_2|$ for the like charges. The singlet state thus lies above the triplet state in energy. The calculated values of the low lying states for the He atom have been depicted in Figure 19. These levels are in excellent agreement with the corresponding experimental values determined from the optical spectrum of He.

One notes that the energy levels calculated above still have degeneracy corresponding to the value of S^2. This degeneracy can be removed by adding the spin-orbit interaction to the Hamiltonian of the system and calculating the refined energy values. Experimentally, such spin-orbit splittings have been observed in He.

For an explicit example of Equation 244 we write the simplified expression for the energy for $n, \ell = 1,0$ (i.e., the 1s state) and $n', \ell' = 2,1$ (i.e., the 2p state):

$$E_{1s,2p} = -2E_oZ\left(\frac{5}{8}Z - \frac{59}{243} \pm \frac{112}{10935}\right) \tag{247}$$

The upper and lower signs in the above equation correspond to the triplet and singlet states, respectively.

There has been considerable improvement[53-70] in the calculated energy values of He and He-like atoms. Many calculations have been based on the variational method[53-58] of quantum mechanics which allows one to assume a trial wave function of a required symmetry with a set of variational parameters and vary the parameters to minimize the energy.

Assuming the simplest trial function of the form

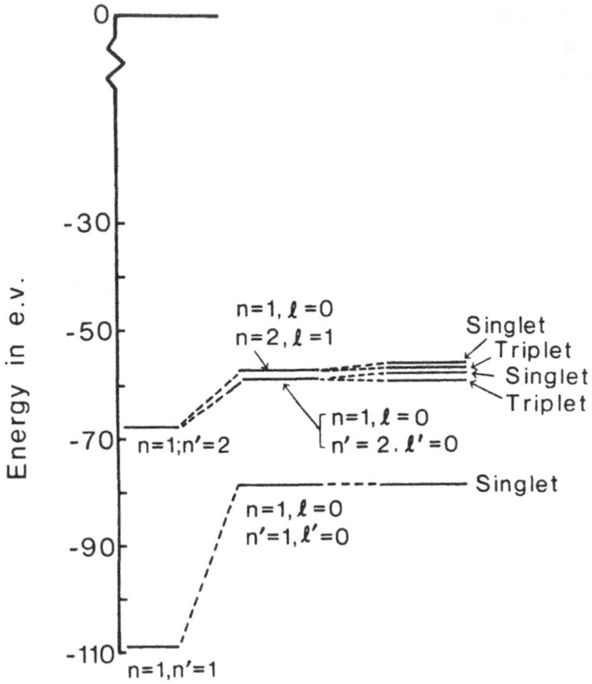

FIGURE 19. Calculated values of the ground state and excited state energy levels; left: if two electrons are not interacting with one another through any field; center: if only the Coulomb interaction J is present; right: if both the Coulomb interaction J and the exchange interaction K are present.

$$\phi \equiv e^{-\alpha(r_1+r_2)} \tag{248a}$$

one obtains for the variational parameter α,

$$\alpha = Z - \frac{5}{16} \tag{248b}$$

and the ground state energy (corresponding to $1s^2$ configuration of He)

$$E = -2E_0\left(Z - \frac{5}{16}\right)^2 \tag{249}$$

where E_0 is defined in Equation 234b.

The calculated results can be improved by including more variational parameters and choosing better forms of the variational function. Thus, Hylleraas[67] and others have used trial functions of the form

$$\phi = A \exp\left[-(Z-\zeta)(r_1 + r_2)\right] \sum_{p',q',r'} C_{p',q',r'} (r_1 + r_2)^{p'} (r_1 - r_2)^{2q'} r_{12}^{r'} \tag{250}$$

Table 5
COMPILATION OF THE CALCULATED GROUND STATE ENERGY OF He ATOM IN UNITS OF $me^4/(\hbar^2 4\pi\varepsilon_o)^2$

Energy	Number of terms or method	Authors and Ref.
−2.75	1	Schwartz,[59] Pekeris,[60,62]
−2.848	1	Schiff et al.,[61] Nesbet and Watson,[63] Tycko et al.,[63] Watson,[64] Bunge,[64] Schaefer and Harris,[64] Green et al.[65]
−2.863	Hartree-Fock	Wilson and Lindsay[66]
−2.90244	3	Hylleraas[67]
−2.90324	6	Hyllerass[67]
−2.903722	38	Kinoshita[68]
−2.90372433	600	Scherr and Knight[69]
−2.903724370	715	Pekeris[70]
−2.903724375	1075	Pekeris[70]
−2.9037243771	164	Schwartz[59]

where A is the normalization constant and p',q',r' are integers; ζ is the screening constant, which is taken to be 5/16 for the ground state. Hylleraas' first variational function contained only three variational parameters with (p',q',r') as (0,0,0), (0,0,1) and (0,1,0), respectively, which were found to be

$$C_{000} = 1, C_{001} = 0.081 \text{ and } C_{010} = 0.010 \tag{251}$$

by minimization of the energy. Pekeris[70] retained a high number of terms, namely 1075, and found the ground state energy with accuracy of 1 in 10^{10}. Schwartz[59] used the same type of function as in Equation 250, but assumed some powers to have half-integer values. He found that one needs only 164 terms to exceed the accuracy of Pekeris. The calculated results for the He atom have been compiled in Table 5 for easy reference.

The variational method has also been employed for the calculations of the excited states of the He atom,[60,61] Li^+ ion,[62] the H^- ion,[62] the lowest singlet state of the Li^+ ion,[61] and the ground state of all He-type ions up to Ne(IX).

E. The Periodic Table and the Electronic Arrangement in Atoms

We have already learnt about the nature of the motion of a single electron around an atomic nucleus and the Pauli exclusion principle. We are now ready to treat two of the most interesting phenomena of atomic physics, namely, the atomic shell structure and the periodic system of elements. Quantum mechanics has been very successful in interpreting the periodic table and has explained the chemical similarities of various groups of elements.

In 1869 Mendeleev pointed out that a periodic relation exists between the chemical properties of an element and its atomic weight. He was able to predict many new elements and their chemical and physical properties. For instance, gallium and germanium were predicted by him and were later discovered with their expected chemical and physical properties. As more new elements were discovered and their chemical properties known, this periodic table was modified. The modern form of the periodic table is shown in Table 6, which consists of periods separated by noble gases. Each

Table 6
THE PERIODIC SYSTEM OF THE ELEMENTS

1st per	1 H																	2 He
2nd	3 Li	4 Be											5 B	6 C	7 N	8 O	9 F	10 Ne
3rd	11 Na	12 Mg											13 Al	14 Si	15 P	16 S	17 Cl	18 Ar
4th	19 K	20 Ca	21 Sc	22 Ti	23 V	24 Cr	25 Mn	26 Fe	27 Co	28 Ni	29 Cu	30 Zn	31 Ga	32 Ge	33 As	34 Se	35 Br	36 Kr
5th	37 Rb	38 Sr	39 Y	40 Zr	41 Nb	42 Mo	43 Tc	44 Ru	45 Rh	46 Pd	47 Ag	48 Cd	49 In	50 Sn	51 Sb	52 Te	53 I	54 Xe
6th	55 Cs	56 Ba	57-71 *	72 Hf	73 Ta	74 W	75 Re	76 Os	77 Ir	78 Pt	79 Au	80 Hg	81 Tl	82 Pb	83 Bi	84 Po	85 At	86 Rn
7th	87 Fr	88 Ra	89 Ac	90 †														

57 La	58 Ce	59 Pr	60 Nd	61 Pm	62 Sm	63 Eu	64 Gd	65 Tb	66 Dy	67 Ho	68 Er	69 Tm	70 Yb	71 Lu	*Rare earths
	90 Th	91 Pa	92 U	93 NP	94 Pu	95 Am	96 Cm	97 Bk	98 Cf	99 Es	100 Fm	101 Md	102 No	103 Lw	†Heavy elements

element is assigned a chemical symbol with its atomic number Z. Groups of elements having similar chemical properties have been put in vertical columns in the periodic table. For example, alkali metals, alkaline earths, halogens, and noble gas elements occupy the vertical columns. The alkalis have been put in the first column since they have a valence of plus 1, the noble gases have been put in the last column since they have a valence zero, and so on. There are also subgroups such as Cu, Ag, Au, and Zn, Cd, Hg; and others which extend vertically only in part of the columns. The arrows in Table 6 have been used to indicate that the subgroups are related to the main group by chemical properties.

The question arises whether one somehow can account for the arrangement of the atoms in the periodic table. In fact, the Bohr theory combined with the Pauli exclusion principle provides a qualitative explanation of the periodicity of the atoms. However, Schrödinger's theory is needed for a quantitative treatment.

We have already seen that the hydrogen eigenfunctions $\psi_{n\ell m_\ell}(\vec{r})$ (see Section II) are characterized by the quantum numbers n, ℓ, m_ℓ. Considering that the states are given by various values of n, ℓ, m_ℓ and applying the Pauli principle, one obtains an approximate scheme of describing the electronic structure of atoms that is found to be very convenient and leads to the shell structure of atoms. According to this simplified model the ground state of the atom can be obtained by placing its two electrons in the lowest state $n = 1$, $\ell = 0$ which we denote by the configuration $(1s)^2$, where the superscript denotes the number of electrons and 1s corresponds to $n = 1$, $\ell = 0$. On the same basis, the configuration of the excited state of the atom can be represented by (1s)(2s), (1s)(3s), (1s,2p), etc., where s,p..., etc. designate the $\ell = 0, 1, \ldots$, etc. states, respectively, (see also Table 7) and the numbers preceding the symbols are the principal quantum numbers. Thus (2p) stands for the $n = 2$ and $\ell = 1$ state.

Table 7
SPECTROSCOPIC NOTATIONS CORRESPONDING TO DIFFERENT ℓ VALUES

ℓ	0	1	2	3	4	5	6
Spectroscopic symbol	s	p	d	f	g	h	i

Table 8
GROUND STATE CONFIGURATIONS

1. H (1s)	19. K...$(3s)^2(3p)^6(4s)$
2. He $(1s)^2$	20. Ca...$(3s)^2(3p)^6(4s)^2$
3. Li $(1s)^2(2s)$	21. Sc...$(3s)^2(3p)^6(3d)(4s)^2$
4. Be $(1s)^2(2s)^2$	22. Ti...$(3s)^2(3p)^6(3d)^2(4s)^2$
5. B $(1s)^2(2s)^2(2p)$	23. V...$(3s)^2(3p)^6(3d)^3(4s)^2$
6. C...$(2s)^2(2p)^2$	24. Cr...$(3s)^2(3p)^6(3d)^5(4s)$
7. N...$(2s)^2(2p)^3$	25. Mn...$(3s)^2(3p)^6(3d)^5(4s)^2$
8. O...$(2s)^2(2p)^4$	26. Fe...$(3s)^2(3p)^6(3d)^6(4s)^2$
9. F...$(2s)^2(2p)^5$	27. Co...$(3s)^2(3p)^6(3d)^7(4s)^2$
10. Ne...$(2s)^2(2p)^6$	28. Ni...$(3s)^2(3p)^6(3d)^8(4s)^2$
11. Na...$(2s)^2(2p)^6(3s)$	29. Cu...$(3s)^2(3p)^6(3d)^{10}(4s)$
12. Mg...$(2s)^2(2p)^6(3s)^2$	30. Zn...$(3s)^2(3p)^6(3d)^{10}(4s)^2$
13. Al...$(3s)^2(3p)$	31. Ga...$(4s)^2(4p)$
14. Si...$(3s)^2(3p)^2$	32. Ge...$(4s)^2(4p)^2$
15. P...$(3s)^2(3p)^3$	33. As...$(4s)^2(4p)^3$
16. S...$(3s)^2(3p)^4$	34. Se...$(4s)^2(4p)^4$
17. Cl...$(3s)^2(3p)^5$	35. Br...$(4s)^2(4p)^5$
18. Ar...$(3s)^2(3p)^6$	36. Kr...$(4s)^2(4p)^6$

In the Li atom there are three electrons and, because of the Pauli exclusion principle, it is not possible to put all three electrons in the 1s state. The ground state of Li can then be the configuration $(1s)^2(2s)$ or $(1s)^2(2p)$ where the two electrons in the same state $(1s)^2$ have opposite spins. From the atomic spectra and from theoretical calculations it is known that the energy of the ns state is always lower than the np state. Accordingly, the ground state of the Li atom possesses the $(1s)^2(2s)$ configuration; $(1s)^2(2p)$ is its excited state.

Similarly, one represents the ground state energy of the Be atom as $(1s)^2(2s)^2$. We list the ground state configurations[71,72] of some of the elements in Table 8. As mentioned earlier, the ns states lie lower in energy than the np states. Analogously, the np states have lower energies than the nd states, and so on. As we go towards the heavier atoms it happens that the 4s states have lower energies than the 3d states and 5s states have lower energies than 4f states. Thus the ℓ-dependence of the energy of the outer subshells can be more predominant than the n-dependence when the principal quantum number n is large. The ordering of the energy levels can be summarized by the following general rule: an outer subshell with lowest ℓ has the lowest energy for a given value of n. If ℓ is fixed, the outer subshell with the lowest n has the lowest energy.

To obtain exact information about the energy ordering of all the subshells in an atom is a difficult task since it requires solving the Schrödinger equation for multielectron atoms. In the next section we shall briefly describe the important methods that have been adopted to solve the Schrödinger equation in such complicated cases. To reveal the complexity of the relative ordering we mention the case of the relative ordering of 3d and 4s subshells; for K and for the next few atoms, the 4s subshell is lower

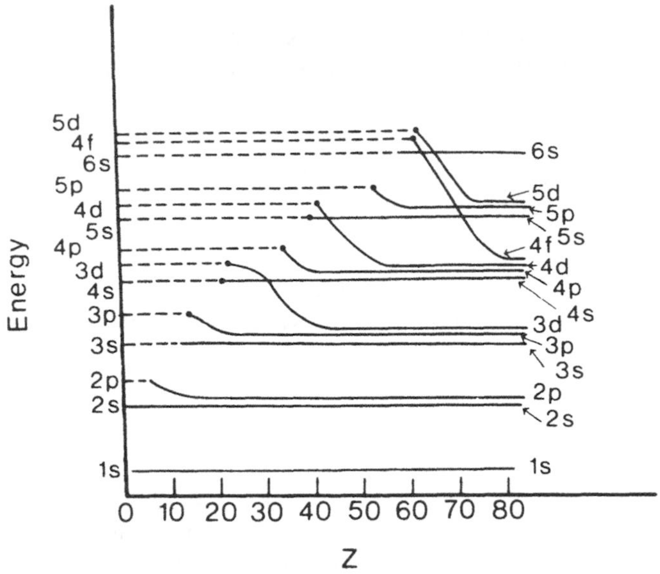

FIGURE 20. Ordering of the subshells as a function of the atomic number Z.

Table 9
SPECTROSCOPIC NOTATIONS FOR VARIOUS SHELLS AND SUBSHELLS AND THE NUMBER OF ELECTRONS THE SUBSHELLS CAN OCCUPY

Shell	K	L		M			N				O					P		
n	1	2		3			4				5					6		
Subshell	1s	2s	2p	3s	3p	3d	4s	4p	4d	4f	5s	5p	5d	5f	5g	6s	6p	6d
ℓ	0	0	1	0	1	2	0	1	2	3	0	1	2	3	4	0	1	2
No. of electrons	2	2	6	2	6	10	2	6	10	14	2	6	10	14	18	2	6	10

than the 3d subshell. This trend is reversed for atoms of higher atomic number. Figure 20 illustrates the ordering of the subshells as a function of the atomic number. Sternheimer[73-77] has recently dealt with the problem of the ordering of atomic energy levels.

Thus we see that the electrons occupy various shells in an atom. Stoner has symbolized the shells and his notations (K-shell, L-shell, M-shell, etc.) have been listed in Table 9 along with the equivalent principal quantum number n. The shells are further divided into subshells characterized by the azimuthal quantum number ℓ. In Table 9 we have also described the subshells and have listed the total number of electrons which can occupy a subshell. Table 10 lists the shell structure of the noble gases separately to show which shells are completely filled in these cases.

As is clear, the interpretation of the periodic table from quantum mechanical considerations is based on the outermost filled subshells of multielectron atoms. The chemical and physical properties of an atom depend on the ground state electronic structure and the character and position of the excited levels. In the case of noble gas elements $_{10}$Ne, $_{18}$A, $_{36}$Kr, $_{54}$Xe, and $_{86}$Rn, the outermost filled subshell is p and the next vacant subshell, which happens to be the s shell, is much higher in energy. Consequently, the noble gas atoms are difficult to excite. Also, the shells of the atoms are completely filled. On account of such characteristics these atoms are inert in their interactions

Table 10
SHELL STRUCTURE OF THE NOBLE GASES

Subshells	Total no. of electrons	Element
1s,	2	He
1s, 2s, 2p	10	Ne
1s, 2s, 2p, 3s, 3p	18	Ar
1s, 2s, 2p, 3s, 3p, 4s, 3d, 4p	36	Kr
1s, 2s, 2p, 3s, 3p, 4s, 3d, 4p, 5s, 4d, 5p	54	Xe
1s, 2s, 2p, 3s, 3p, 4s, 3d, 4p, 5s, 4d, 5p, 6s, 4f, 5d, 6p	86	Em

with other atoms to produce chemical compounds. Furthermore, they have very low freezing and boiling points since their atoms do not have any appreciable tendency to interact with one another.

From the electronic structure it is easy to understand quantitatively the ionization energies of atoms. The ionization energy is equal to the energy required to remove the electron from the atom and thus depends on the energy of the outermost filled level. The ionization energies of the alkalis $_3$Li, $_{11}$Na, $_{19}$K, $_{37}$Rb, $_{55}$Cs, and $_{87}$Fr are particularly small since the outermost-bound electron is in the s shell, which is a weakly bound state. They are also chemically active since it is easy (energetically) for them to release the outermost electron. Because of the single active electron, these atoms are said to have a valence of plus one and are known as univalent atoms.

On the other hand, the halogens $_9$F, $_{17}$Cl, $_{35}$Br, $_{53}$I, and $_{85}$At have a high electron affinity since their outermost shell (which is a p shell) has a vacancy of one electron. Thus they have a valence of minus one.

The atoms $_{21}$Sc through $_{28}$Ni are known as the first-transition group elements. They have similar properties since they contain 3d electrons. Because of the directional properties of the d shells (see Figure 8), these play important roles in catalysis; many of these elements and their compounds are known to have good catalytic properties in chemical reactions.

In the case of the rare earth elements $_{58}$Ce through $_{71}$Lu, the 4f shells start being filled. The 4f shells lie within the filled 6s shells and therefore are completely shielded from the external environment. Owing to this, they possess similar chemical properties. Analogously, actinides, $_{90}$Th through $_{103}$Lw, have 5f partly filled subshells and are shielded by 7s shells.

F. Various Theoretical Methods

We wish to give here a brief account of the quantum mechanical study of multielectron atoms. As compared to the one-electron atom, the treatment for multielectron atoms is very complicated. However, the efforts expended on these systems are quite worthwhile since the results derived from such (quantum mechanical) studies provide a detailed understanding of the atoms which are, no doubt, constituents of almost everything in the universe.

A complex atom or ion is composed of a nucleus of charge Ze and N electrons each of charge (−e). Neglecting several relativistic and spin-dependent effects, one writes the Hamiltonian operator of the system as

$$\mathcal{H} = -\frac{\hbar^2}{2m}\sum_{i}^{N}\nabla_1^2 \sum_{i-1}^{N}\frac{Ze^2}{(4\pi\epsilon_o)r_i} + \sum_{i>j}\frac{e^2}{(4\pi\epsilon_o)\,r_{ij}} \qquad (252)$$

where the first term in $\mathcal{H}$ is the operator corresponding to the sum of the kinetic energies of all the electrons, the second term is the Coulomb potential energy of interaction between all the electrons and the nucleus, and the third term is the Coulomb interaction energy between all the electron pairs.

For obtaining the eigenvalues and eigenfunctions of the Hamiltonian in Equation 252 one is required to solve the time-independent Schrödinger equation

$$\mathcal{H}\Psi = E\Psi \qquad (253)$$

where Ψ depends on the coordinates of all the N electrons and E is the total energy of the system. It is impossible to solve the Schrödinger Equation 253 exactly for the multielectron atom. Thus one has to resort to approximate treatments. Fortunately, we are guided by the solutions of the Schrödinger equation for one-electron atoms and the experimental information from optical spectra and chemical properties and, to some extent, by the experience gained in calculations performed on the He atom.

In the following we shall describe various theories adopted for obtaining quantitative information for the eigenvalues and eigenfunctions of the Schrödinger Equation 253.

1. The Hartree Self-Consistent Field Theory

Hartree,[78,79] in 1928, proposed a method applicable to multielectron atoms. In this method one assumes in the first approximation that (1) the atomic electrons move independently of one another so that the motion of one electron does not depend on the motion of the other electron, (2) each electron moves in a spherically symmetric net potential V(r) produced by the spherically symmetric attractive Coulomb potential due to the nucleus, and a spherically symmetric repulsive potential from the average effect of the repulsive Coulomb interactions arising from its N − 1 partners, and (3) the weaker condition, viz., the Pauli exclusion principle (instead of the stronger condition which demands the antisymmetry of the multielectron wave function with respect to the exchange of the coordinates, including spins, of any two electrons) is satisfied by requiring that only one electron populates each quantum state.

For a spherically symmetric central potential V(r), Hartree wrote down the time-independent Schrödinger equation for the ith electron in the form

$$\left[-\frac{\hbar^2}{2m}\nabla_i^2 + V(r_i)\right]\Psi(r_i, \theta_i, \phi_i) = \epsilon\Psi(r_i, \theta_i, \phi_i) \qquad (254)$$

where r_i, θ_i, and ϕ_i designate the spherical polar coordinates (with respect to the nucleus) of the ith electron, and ε is the total energy for the ith electron with the net potential $V(r_i)$ and the eigenfunction $\Psi(r_i,\theta_i,\phi_i)$.

According to the original treatment of Hartree, it is assumed at the onset that the potential V(r) very close to the center of the atoms is essentially due to the nuclear charge Ze, and very far from the center V(r) is due to a nucleus of charge e (Z − N + 1), which represents the nuclear charge Ze shielded by the charge −e (N − 1) of the (N − 1) other electrons, with a reasonable interpolation of V(r) for the intermediate values of r. Thus, initially one assumes,

$$V(r) = \begin{cases} \dfrac{-Ze^2}{4\pi\epsilon_0 r} & \text{for } r \to 0 \\[2ex] \dfrac{(Z-N+1)\,e^2}{4\pi\epsilon_0 r} & \text{for } r \to \infty \end{cases}$$

with some interpolation for V(r) for $0 < r < \infty$. Hartree solved the Schrödinger Equation 254 with this initial V(r) and obtained $\psi(r_i,\theta_i,\phi_i)$ and the corresponding ε for all the N electrons. The quantum states were filled with the electrons so as to minimize the total energy, consistent with the Pauli exclusion principle. Next, Gauss's law of electrostatics was employed to calculate the potential at any point r by taking the charge distributions $-e\Psi^*(r_j\theta_j\phi_j)\Psi(r_j\theta_j\phi_j)$ of all the other (N − 1) electrons; the corrected new potential V(r) was obtained by adding to it the potential due to the nuclear charge Ze. This potential was then used in Equation 254 to derive the new sets of $\psi(r_i,\theta_i,\phi_i)$ and the corresponding ε. The process was repeated until the self-consistency was attained, that is, the potential V(r) obtained at the end of a cycle was essentially the same as that used in the beginning.

One notes that because of the centrally symmetric nature of the potential, the Schrödinger Equation 254 has a solution of the form (see Section II)

$$\Psi(r_i, \theta_i, \phi_i) \equiv \Psi_{n\ell m_\ell}(r_i, \theta_i, \phi_i) = R_{n\ell}(r) Y_\ell^{m_\ell}(\theta_i, \phi_i) \tag{255}$$

where $Y_\ell^{m_\ell}(\theta_i\phi_i)$ are the spherical harmonics and $R_{n\ell}(r)$ are the radial functions which satisfy the equation

$$\left[-\frac{\hbar^2}{2m}\left\{\frac{1}{r_i^2}\frac{\partial}{\partial r_i} r_i^2 \frac{\partial}{\partial r_i} - \frac{\ell_i(\ell_i+1)}{r_i^2}\right\} + V(r_i)\right] R_{n\ell}(r_i)$$

$$= \epsilon_{n_i\ell_i} R_{n_i\ell_i}(r_i) \tag{256}$$

or equivalently,

$$\left[-\frac{\hbar^2}{2m}\left\{\frac{d^2}{dr_i^2} - \frac{\ell_i(\ell_i+1)}{r_i^2}\right\} + V(r_i)\right]$$

$$P_{n_i\ell_i}(r_i) = \epsilon_{n_i\ell_i} P_{n_i\ell_i}(r_i) \tag{257}$$

where

$$P_{n_i\ell_i}(r_i) = r_i R_{n_i\ell_i}(r_i) \tag{258}$$

In the above equations we have used the subscript $n_i\ell_i$ on ε, R(r) and P(r) to express the fact that these quantities depend explicitly on the quantum numbers $n_i\ell_i$ of the ith electron.

It is then clear that in the Hartree approximation the quantum numbers $n_i\ell_i m_l$ are the same as in the case of a single-electron atom (Section II) and give rise to the shells and subshells of the electrons in atoms. The wave functions $R_{n_i\ell_i}(r_i)$ and the energies $\epsilon_{n_i\ell_i}$ are no longer the same as for the one-electron atom. This is because the potential $V(r_i)$ is not a simple Coulomb potential of the nuclear charge alone, but also contains the average spherically symmetric potential (different for electrons of different quantum numbers) due to the other electrons. The Hartree theory not only gives rise to the shell structure for the multielectron atoms and ions but also forms the basis for the quantum mechanical explanation of the periodic table of the elements elaborated in the preceding subsection.

It is possible to obtain the Hartree equations directly by applying the variational principle in which the energy of the system is minimized with respect to the one-electron orbitals[10,80] $\psi_{n_i \ell_i m_{\ell i}}$. The N electron total wave function is assumed to be of the product form,

$$\Psi(1, 2, \ldots N) = \psi_{n_1 \ell_1 m_1}(r_1 \theta_1 \phi_1) \cdots \psi_{m_n \ell_n m_n}(r_n \theta_n \phi_n) \tag{259}$$

which does not contain the spin functions explicitly, but the Pauli exclusion principle is followed by assigning no more than two electrons to a given state ($n\ell m_\ell$) of which one corresponds to $m_s = \frac{1}{2}$ and the other to $m_s = -\frac{1}{2}$. In Equation 259 no assumptions have been made regarding the variational form of the one-electron wave functions $\psi_{n\ell m_\ell}$.

Now define the energy-integral

$$E = \langle \Psi | \mathcal{H} | \Psi \rangle \tag{260}$$

where $\mathcal{H}$ and Ψ are given by Equations 252 and 259, respectively. The auxiliary restrictions on the one-electron orbitals are that they are normalized. Thus,

$$\langle \psi_{n_i \ell_i m_{\ell_i}} | \psi_{n_i \ell_i m_{\ell_i}} \rangle = 1 \tag{261}$$

One now carries out the variation of E (Equation 260) with the conditions of Equation 261 following the method of undetermined multipliers. For minimization of the energy E we have

$$\delta \left[E - \Sigma_i \epsilon_{n_i \ell_i} \langle \psi_{n_i \ell_i m_i} | \psi_{n_i \ell_i m_i} \rangle \right] = 0 \tag{262}$$

where $\epsilon_{n_i \ell_i}$ are the undetermined multipliers. Since no functional form of $\psi_{n_i \ell_i m_i}$ is assumed, the variation $\delta \psi_{n_i \ell_i m_i}(r_i)$ is independent of $\vec{r}_i$. One assumes also that the variation of one orbital function is independent of the other. The variation leads to the Hartree equations.

$$\left[\frac{-\hbar^2}{2m} \left\{ \frac{d^2}{dr_i^2} - \frac{\ell_i(\ell_i + 1)}{r_i^2} \right\} - \frac{Ze^2}{4\pi\epsilon_o r_i} + \frac{e^2}{4\pi\epsilon_o} \frac{1}{r_i} \sum_{j \neq i} \quad P_{n_j \ell_j}(r_j) \frac{r_i}{r_>} P_{n_j \ell_j}(r_j)\, dr_j \right] P_{n_i \ell_i}(r_i)$$

$$= \epsilon_{n_i \ell_i} P_{n_i \ell_i}(r_i) \tag{263}$$

where $P_{n_i \ell_i}(r_i)$ is r times the radial wave function as in Equation 258 and

$$r_> = \begin{cases} r_i & \text{if } r_i > r_j \\ r_j & \text{if } r_j > r_i \end{cases} \tag{264}$$

In arriving at Equation 263 the spherical average of the electron-electron interaction

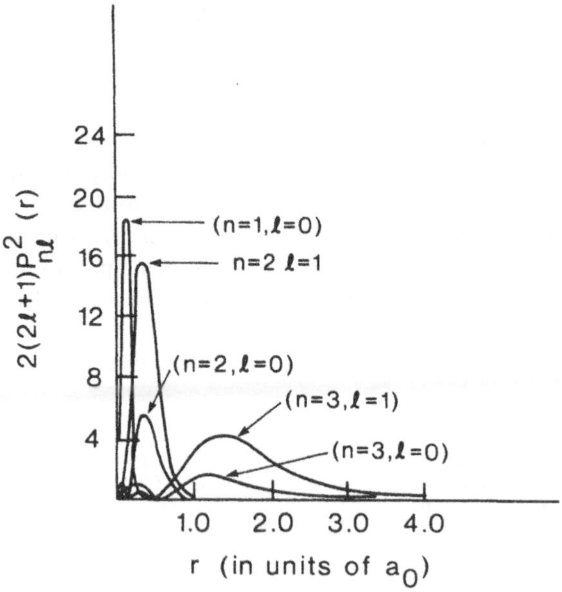

FIGURE 21. Radial probability densities as calculated by the Hartree theory for the argon atom.

has been taken.[10] Equation 263 is exactly the same as Equation 257 with the added advantage that $V(r_i)$ is now explicitly expressed as

$$V(r_i) = -\frac{Z(r_i)\,e^2}{4\pi\epsilon_0}\,\frac{1}{r_i} \qquad (265)$$

where

$$Z(r_i) = Z - \sum_{j\neq i} \int P_{n_j\ell_j}(r_j)\,\frac{r_i}{r_>}\,P_{n_j\ell_j}(r_j)\,dr_j \qquad (266)$$

In the Hartree Equation 263 it is exhibited that a slightly different radial Schrödinger equation is solved for each electron orbital with the potential $V(r_i)$ due to the nucleus and the spherical averaged potential of all orbitals except the one in question. Consequently, any two solutions $P_{n_i\ell_i}$ are not exactly orthogonal for the same values of ℓ_i (and different n_i) though they are orthogonal for different ℓ_i by virtue of the orthogonality in the angular part of the wave functions, the Y^{m}_{ℓ}. However, the Schmidt orthogonalization technique may be adopted to make two orbitals with the same ℓ_i (and different n_i) orthogonal to one another. In the Hartree-Fock method (see next subsection) this difficulty is not encountered even when each electron moves in a different potential.

The radial probability density $2(2\ell + 1)\,P^2_{n\ell}(r)$ for the occupied electron quantum states of the argon atom has been depicted in Figure 21 as a function of r (in units of a_0, the Bohr radius). One notes that shell-like behavior is evident because for each n the probability density is largely concentrated in a limited range of r. Also, the outermost shell (corresponding to n = 3) has the characteristic radius only slightly larger than the Bohr radius a_0 whereas the characteristic radius of the innermost shell (n = 1) has a value much smaller than the Bohr radius.

As emphasized above, the electrons move in a restricted region of r space forming shells in the Hartree approximation. Also, Z(r) (see Equation 266), the effective charge, has a definite value in that particular region of shells. From these considerations it is possible to obtain a rough, although useful, description of the multielectron system in terms of a charge Z_n which assumes a constant value equal to the average value of Z(r) at the shell radius of principal quantum number n. Z_n may be called the effective Z for the shell. In this crude approximation it is possible to deduce many approximate, but useful, results. On analysis of the Hartree calculations one observes that in all multielectron atoms Z_n has a value $Z_n \simeq n$ if n is the principal quantum number of the outermost occupied shell. Other results may be summarized as follows:

1. Since $Z_n \simeq n$ for the outermost shell, the approximate energy of the outermost shell is

$$\epsilon \simeq -\frac{\mu Z_n^2 e^4}{(4\pi\epsilon_0)^2 2\hbar^2 n^2}$$

$$= -\frac{\mu e^4}{(2\pi\epsilon_0)^2 2\hbar^2} \tag{267}$$

2. The radius of the outermost electron can analogously be derived as

$$\bar{r} = \frac{n^2 a_0}{Z_n} \simeq \frac{n^2 a_0}{n} \simeq n a_0 \tag{268}$$

We see that the radius of the outermost shell increases slowly with increasing atomic number. This explains the fact that atoms of high atomic number are not very much bigger than atoms of low atomic number.

3. The radii of shells of small n are very small since the electrons feel the full Coulomb attraction of the positive nucleus. The effective charge for the inner shells is crudely given by

$$Z_1 = Z - 2 \tag{269}$$

which yields for the radii of the inner shells,

$$\bar{r} = \frac{n^2 a_0}{Z_1}$$

$$\simeq \frac{\bar{r}(\text{hydrogen})}{Z - 2} \tag{270}$$

by substituting $Z_1 = Z - 2$.

4. Similarly, the energy of the inner shell orbitals for the multielectron atom is given by

$$\epsilon = -\frac{\mu Z_1^2 e^4}{(4\pi\epsilon_0)^2 2\hbar^2 n^2}$$

$$\simeq E(\text{hydrogen})\ (Z - 2)^2 \tag{271}$$

5. In the Hartree approximation one also obtains

$$\psi_{n\ell m_\ell} \propto r^\ell \quad (\text{for } r \to 0)$$

as can be proved from Equation 263, which is true for any potential V(r) that does not increase in magnitude as fast as $1/r^2$ as r tends to zero. Accordingly, the probability density for an electron behaves as

$$\psi^*\psi \propto r^{2\ell} \quad (\text{for } r \to 0)$$

which explains the high probability for the s-state electrons to be located near the nucleus, a result of high significance for explaining the hyperfine structure of atoms.

2. The Hartree-Fock Self-Consistent Field Theory

We have seen in the preceding section that in the Hartree method one sets up a product of one-electron orbitals of various electrons, finds the average value of the Hamiltonian, and minimizes the energy by varying the one-electron orbitals. However, such a product function does not satisfy the antisymmetry principle of Dirac, though the weak condition, namely the Pauli principle, is satisfied. Naturally, if one assumes the proper antisymmetrized form of the wave function, one expects to obtain more accurate results than the Hartree results. Fock[81] originally suggested the use of antisymmetrized wave functions. The simplest function that satisfies the antisymmetry condition is a determinantal function or a linear combinations of determinantal functions. Assuming such a form of the function and minimizing the energy with respect to the one-electron orbitals constitute the Hartree-Fock method.[79] An example where one can use a single determinantal function is that of the system of closed shells, such as inert gas atoms or complete ion-filled shells. We plan to give here a brief account of the Hartree-Fock theory only for the case pertaining to a single determinantal function.

For the N-electron system for which the Hamiltonian is given by Equation 252, the multielectron determinantal function can be written as a Slater determinant,

$$\Psi(1, 2, 3, \ldots N) = (N!)^{-\frac{1}{2}} \begin{vmatrix} u_1(1) & u_1(2) \ldots u_1(N) \\ u_2(1) & u_2(2) \ldots u_2(N) \\ \ldots & \ldots \\ u_N(1) & u_N(2) \ldots u_N(N) \end{vmatrix} \quad (272)$$

where $u_i(j)$ are the one-electron spin orbitals assumed to be the product of a space-dependent function and a function of α or β of spin; the spin-orbitals are appropriate to a definite value of m_s. In $u_i(j)$, j stands for the spatial as well as spin coordinates of an electron whereas i distinguishes different forms of the orbitals. We assume that (1) all the spatial orbitals are normalized, (2) all the spatial orbitals corresponding to $m_s = ½$ are orthogonal to each other, and (3) all the spatial orbitals corresponding to $m_s = -½$ are orthogonal to each other. One then obtains the expectation value of the Hamiltonian (Equation 252)

$$< \Psi | \mathcal{H} | \Psi > = \sum_i < u_i (1) | f_1 | u_i (1) >$$

$$+ \sum_{i \neq j} [< u_i (1) \; u_j (2) | g_{12} | u_i (1) \; u_j (2) >$$

$$- \delta (m_{s_i,} \; m_{s_j}) < u_i (1) \; u_j (2) | g_{12} | u_i (2) \; u_j (1) >] \quad (273)$$

where f_i is the one-electron operator

$$f_i = - \frac{\hbar^2}{2m} \nabla_i^2 - \frac{Ze^2}{4\pi\epsilon_o r_i} \quad (274)$$

and g_{12} is the two-electron operator

$$g_{12} = \frac{e^2}{4\pi\epsilon_o r_{12}} \quad (275)$$

which occur in the Hamiltonian Equation 252.

By varying the one-electron orbitals to minimize the energy (Equation 273) with the restriction that the one-electron orbitals are normalized, i.e.,

$$< u_i | u_j > = \delta_{i,j} \quad (276)$$

one obtains the following Hartree-Fock equation

$$f_1 u_i (1) + \sum_j [< u_j (2) | g_{12} | u_j (2) > u_i (1)$$

$$- \delta(m_{s_i,} \; m_{s_j}) < u_j (2) \; | g_{12} | u_i (2) > u_j (1)] = \epsilon_i u_i (1)$$

$$(277)$$

On comparison with the Hartree equations (see Equation 263), one finds that in the Hartree-Fock treatment some more terms are contained on the left-hand side of Equation 277 which are referred to as the exchange terms.

The quantity ε_i appearing in Equation 277 represents the negative of the energy required to remove the ith electron from the atom provided that the electron orbitals of the ion remain the same as for the atom. This is known as the Koopman theorem[82] and can be easily proved starting with Equation 277.

Next we turn to understanding the physical meaning of the Hartree-Fock equation, particularly the exchange terms. To this end we rewrite Equation 277 as follows:

$$f_1 u_i (1) + \sum_j \left[< u_j (2) | g_{12} | u_j (2) > \right.$$

$$\left. - \delta (m_{s_i,} \; m_{s_j}) \frac{u_i^*(1) < u_j (2) | g_{12} | u_i (2) > u_j (1)}{u_i^* (1) \; u_i (1)} \right] u_i (1)$$

$$= \epsilon_i u_i (1) \quad (278)$$

The last term, involving the exchange terms on the left-hand side of Equation 278 is really a correction term that requires interpretation. This correction term may be regarded as the potential energy at electron 1 due to a charge distribution at the location of electron 2, of value

$$\rho = -\sum_j \delta\,(m_{s_i}, m_{s_j})\,\frac{u_i^*(1)\;u_j^*(2)\;u_j(1)\;u_i(2)}{u_i^*(1)\;u_i(1)} \qquad (279)$$

which is usually referred to as the exchange charge density. The exchange charge density (because of the occurrence of the Krönecker function $\delta(m_{s_i}, m_{s_j})$ in Equation 279) consists of the electron charge densities having the same spin as the spin-orbital u_i, which one finds by solving Equation 278. If point 2 is identical with point 1, the exchange charge density reduces to

$$\rho - \Sigma_j\,\delta(m_{s_i}\; m_{s_j})\; u_j^*(1)\; u_j(1)$$

which is equal in magnitude to the net charge density of all electrons of the same spin as the ith one, at the location 1. Also, the total amount of ϱ in Equation 279 can be proved[10] to be equal to one electronic charge if u_i is one of the occupied spin-orbitals, and equal to zero if u_i is an excited orbital. This is consistent with the physical concept that an occupied spin-orbital will experience the field of the nucleus and the other $N - 1$ electrons, whereas an excited spin-orbital is a solution of the Schrödinger equation for a hypothetical charge which is really zero and is influenced by all N electrons in the occupied spin-orbitals.

Because of the exchange terms there is a lowering of the potential energy in the neighborhood of the nucleus for an electron in an atom. Thus we have a deeper potential minimum than we have in the Hartree method. As a result, the Hartree-Fock wave functions are such that they lead to the concentration of the charge density more towards the nucleus. Also, one infers that the one-electron energy parameters ε_i are lower in the Hartree-Fock method than in the Hartree method. This is because in the Hartree method, for the occupied orbital u_i, the charge density of the electrons of the same spin is not only completely missing (owing to the antisymmetry of the wave function) from the immediate neighborhood of the electron in question, but also remains much further away. As a result, the Coulomb repulsion between the electrons is very much reduced in the Hartree-Fock treatment, giving ε_i lower than in the Hartree treatment.

In actual atomic calculations the spatial function of the spin-orbital is usually taken to be in the form of Equation 255. The radial functions $R_{n\ell}(r)$ are calculated numerically and are given as numerical tables.[83] With the advent of computers it is now possible to assume an analytical form of the $R_{n\ell}(r)$ taken as a linear combination of Slater-type orbitals (STOs). Accordingly,

$$R_n(r) = \sum_i C_i r^{n_i - 1} \exp(-\zeta_i r) \qquad (280)$$

where C_i and ζ_i are determined variationally. $r^{n-1}\exp(-\zeta r)$ are the STOs where n is a positive integer ($n \neq 0$). Roothaan[84,85] has employed such wave functions for atomic structure calculations and the method is known as the Hartree-Fock-Roothaan (HFR) method. This method is found to be very useful for calculating the electronic wave functions and energies of very complex atomic and ionic systems. The calculations relevant to the d and f orbitals in atoms have been handled by Watson[86] and others.[87-92] There have also been many attempts to combine the Dirac formalism with the HFR method in order to incorporate relativistic effects which are very significant for heavier atoms.[93-98]

Clementi and Raimondi[99] have calculated (using the HFR method) the electronic structure of the ground state of atoms up to Kr. They employed what is known as the minimum basis set which is concerned with retaining the minimum number of terms, that is, N/2 or (N/2 + 1) in the expansion Equation 280, and used STOs to obtain a reasonably good accuracy. Such functions are accurate enough for applications to solids and molecules. Watson has also employed the minimum basis set for his calculations on d-electron systems.

Naturally, the accuracy of the minimum basis set results can be improved by increasing the number of terms in Equation 280. If the number of terms in the set is twice the number of terms in the minimum basis set, the new set is termed the double zeta set. Clementi[100] has utilized such a set for his calculations for atoms up to Ar and found it to be beneficial for molecular and solid state problems.[101,102] Hartree-Fock wave functions for several atoms and ions have also been calculated by Bagus et al.[103] using an expansion of the form given in Equation 280.

There are also approximate methods available for calculating the atomic structure of atoms, which may be regarded as approximations to the Hartree-Fock method. Such methods are the Thomas-Fermi method[104,105] and the Coulomb approximation method.[106,107] In the Thomas-Fermi method the shell structure is lost, but properties such as the averaged electron distributions and ionization energies can be calculated easily. One expects to obtain reasonably good results for heavier atoms and ions by this method. As for the Coulomb approximation method suggested by Bates and Damgaard,[106] one finds that this forms a relatively simple method and yields results that agree well with the Hartree-Fock method, particularly for electric dipole matrix elements. In the Coulomb approximation method one uses the solutions of the hydrogenic Schrödinger equation with the energy parameters obtained through experiments. Explicitly, the wave function used is of the form

$$\psi_{CA} = \left[\frac{n^{*2}\,\Gamma(n^*+\ell-1)\,\Gamma(n^*-1)}{\gamma}\right]^{-1/2}\left(\frac{2r\gamma}{n^*}\right)^{n^*} \left\{\left[1+\sum_{\nu=1}^{\infty}(a_\nu/r^\nu)\right]/r\right\}\exp(-r\gamma/n^*) \tag{281}$$

where

$$a_1 = \frac{n^*}{2\gamma}\,[(\ell(\ell+1) - n^*(n^*-1)] \tag{282a}$$

$$a_\nu = a_{\nu-1}\left\{\frac{n^*}{2\nu\gamma}\,[\ell(\ell+1) - (n^*-\nu)\,(n^*-\nu+1)]\right\} \tag{282b}$$

$$\gamma = \frac{\mu e^2}{4\pi\epsilon_0\hbar^2} \tag{282c}$$

The above function is appropriate for $r \rightarrow \infty$ but it does not behave well at the origin. This drawback is not serious for calculating properties which depend predominantly on the contributions from the wave function arising from the outer regions of the atom. This is particularly true outside the core of the atom where the potential is largely Coulombic.

The Thomas-Fermi and Coulomb approximation methods work well in only some cases. The latter method is appropriate particularly when one electron is excited to an outer shell. It is, however, a poor representation for an electron in a shell containing more than one electron. Armstrong and Purdum[108] have concluded that for calculating the transition probabilities from a ground state to an excited state it is convenient and accurate enough to use the Coulomb approximation wave function for the excited state and the Hartree-Fock wave function for the ground state.

It is obvious that the Hartree-Fock method is a great advancement over the Hartree method. In Hartree's original method of self-consistent fields, the correlation between the motion of electrons is absent since each electron moves in an average field of other electrons and the wave function is the product of functions of the coordinates of single electrons, thereby making the electron motions statistically independent. The Hartree-Fock method, on the other hand, introduces correlation between electrons of the same spin; two electrons of the same spin are kept apart by virtue of the antisymmetric wave functions used. However, in the Hartree-Fock method no correlations between electrons of the opposite spin have been possible.

3. Hartree-Fock Theory with Slater Exchange

In the Hartree-Fock method one is required to solve a complicated set of simultaneous equations and, for maintaining self-consistency, one has to carry out a tedious iteration process. Not only does one solve a different differential equation for each orbital, but one computes a different exchange potential for each orbital. However, the exchange potential involves a sum of many exchange integrals, depending on the number of electrons involved; the evaluation of the exchange potential becomes more and more difficult as the number of electrons increases. Thus one looks for a simpler method that not only retains the characteristics of the Hartree-Fock method but also facilitates the calculations.

For this purpose Slater[109] has suggested a method which is concerned with replacing the exchange potentials for different orbitals in the Hartree-Fock equations by a universal exchange potential given by

$$V_{ex}(\vec{r}) = \frac{-6\,e^2}{4\pi\epsilon_o}\left[\frac{3}{8\pi}\,|\rho(r)|\right]^{1/3} \qquad (283)$$

where ϱ is the total electronic charge density (both spins). In writing Equation 283 one assumes that there are equal numbers of "up" and "down" spins present.

The above form of the exchange potential eliminates completely the calculations of the exchange integrals. The physical origin of the Slater exchange potential (Equation 283) is contained in the fact that the Hartree-Fock exchange charge density (Equation 279) is of opposite sign (but with the same spin orientation) of the charge density of the electron in question. This turns out to be equivalent to forming a hole (known as the exchange hole or Fermi hole) which carries a total charge in magnitude equal to one electron charge. If R is the radius of the hole, $4/3\ \pi R^3\varrho$ must be equal to one electron charge, which leads to a radius of the hole proportional to $1/\varrho^{1/3}$ (assuming that ϱ remains constant throughout the hole, which is strictly valid for a free-electron charge distribution). Now, since the potential at the center of a uniformly charged sphere is proportional to 1/R we obtain the essential result of Equation 283 that the exchange potential is proportional to $\varrho^{1/3}$.

It is evident from the above discussion that the exchange potential in Equation 283 is really the free-electron exchange potential. In applying this expression to atomic systems one specifically assumes that the average exchange potential at point $\vec{r}$ is equal to the exchange potential for a free-electron gas whose total electron-charge density, including both spins, matches that of the atomic system in question.

As mentioned earlier, the exchange charge density has a total charge corresponding to one electron charge, and its value at the position of the electron (with which it interacts) is equal in magnitude to the total charge density due to all the electrons possessing the same spin as this electron. For the sake of completeness we also mention Maslen's conclusions[110] which point out that (1) the exchange hole in atomic systems is, in general, nonspherical, (2) the exchange hole does not follow the electron on which it acts and is not localized in the neighborhood of this electron but lags behind, and (3) the maximum value of the exchange hole does not lie at the location of the electron. Maslen's study therefore fixes some limitations on the use of the Slater exchange potential for atomic calculations. For further details on the exchange approximation we refer to articles by various authors.[109,111-115]

Originally Slater[109] was concerned with his exchange approximation for calculations on the closed shell system of atoms, for which he built the many electron wave functions as a single determinantal wave function formed by spin orbitals. Later on, numerous authors[116-122] have extended this approximation to an open shell configuration containing unequal numbers of "up" and "down" spin orbitals. Essentially, they took a single determinantal wave function but assigned different radial orbitals to spin "up" and spin "down" electrons. These studies prove the importance of the exchange polarization. This approximation, however, is unsuitable for investigating the excited states of atoms and is found to violate certain symmetry requirements.

Assuming that the multielectron wave function can be represented by a single determinantal wave function constructed from one-electron spin orbitals, one can write the radial HFS wave equations, applicable for free atoms or ions, as:

$$\left[-\frac{\hbar^2}{2m}\left\{\frac{d^2}{dr^2}-\frac{\ell(\ell+1)}{r^2}\right\}+V(r)\right]P_n(r)=\epsilon_{n\ell}P_{n\ell}(r) \tag{284}$$

where

$$V(r)=-\frac{Ze^2}{4\pi\epsilon_0 r}+\frac{e^2}{4\pi\epsilon_0}\frac{1}{r}\int_0^r\rho(r')\,4\pi r'^2\,dr' + \frac{e^2}{4\pi\epsilon_0}\int_r^\infty\frac{\rho(r')}{r'}4\pi r'^2 dr' - \frac{6\,e^2}{4\pi\epsilon_0}\left[\frac{3}{8\pi}\rho(r)\right]^{1/3} \tag{285}$$

$$P_{n\ell}(r) = r\,R_{n\ell}(r) \tag{286}$$

and $\varrho(r)$ is the (spherical) total electronic charge density for both spins,

$$\rho(r)=\frac{1}{4\pi r^2}\sum_{n\ell}\omega_{n\ell}\,[P_{n\ell}(r)]^2 \tag{287}$$

where $\omega_{n\ell}$ is the occupation number of the orbital $n\ell$ (counting electrons of both spins). Evidently,

$$N=\sum_{n\ell}\omega_{n\ell} \tag{288}$$

Equation 284 is applicable to a single electronic configuration of an atom or ion having one or more open shells.

Table 11
ENERGY LEVELS OF Cu^+ IN RYDBERG UNITS

$n\ell$	Hartree method[125]	Hartree-Fock-Slater method[a] [124]	Hartree-Fock method[126]
1s	−658	−650.40	−685.4
2s	−78.45	−78.87	−82.30
2p	−69.86	−69.74	−71.83
3s	−8.986	−9.355	−10.651
3p	−6.078	−6.429	−7.279
3d	−1.195	−1.459	−1.613

[a] Using V′ (r) of Equation 289.

For large distances the potential V(r) for an atom in Equation 285 is faulty since it approaches zero instead of $-e^2/4\pi\varepsilon_0 r$, which is required because an outer electron at large distances must experience the field of a singly charged positive ion. This situation is rectified by a procedure suggested by Latter.[123] Accordingly, the potential V(r) is replaced by V′(r) where

$$V'(r) = \begin{cases} V(r) \text{ for } r < r_o \\ \dfrac{-e^2}{(4\pi\epsilon_o)} (Z - N+1)/r \text{ for } r \geq r_o \end{cases} \qquad (289)$$

where r_0 is defined by

$$V(r_o) = \frac{-e^2}{(4\pi\epsilon_o)} \frac{(Z-N+1)}{r_o} \qquad (290)$$

The discontinuity in the slope of V′(r) has not been found to induce discontinuities at $r = r_0$ in wave functions or their first and second derivatives.

Making use of the modified potential, V′(r), Herman and Skillman[124] have solved Equation 284 self-consistently to obtain one-electron wave functions for all normal neutral atoms from helium through lawrencium. These wave functions have been demonstrated to be very useful by several investigators for studying numerous properties of atoms, ions, molecules, and solids. To depict the accuracy of the calculations we compare in Table 11 the energy values of various levels of Cu^+ in units of Rydberg (1 Rydberg = 10967757.6 ± $1.2m^{-1}$) as obtained by Herman and Skillman (for the potential V′(r)) with the corresponding values obtained by Hartree[125] and Hartree-Fock methods.[126] With the exception of the energy levels for the 1s and 2p shells, the Herman-Skillman values lie in between the corresponding Hartree and Hartree-Fock values.

For systems more complicated than atoms, Slater[127] and Slater and Johnson[128] have developed a method known as the Xα method, which essentially uses the Slater exchange approximation as explained above.

4. Density-Functional Formalism

The density functional (DF) formalism of Kohn and Sham[129] starts with the proof[130] that the ground state energy E_o of a nonhomogeneous system of interacting electrons in a static external potential $V(\vec{r})$ can be written in the form

$$E = \int V(\vec{r})\,\rho(\vec{r})\,d^3r + \frac{e^2}{4\pi\epsilon_0}\,\frac{1}{2}\iint \frac{\rho(\vec{r})\,\rho(\vec{r}')}{|\vec{r}-\vec{r}'|}\,d^3r'\,d^3r + G(\rho(\vec{r})) \tag{291}$$

where $\varrho(\vec{r})$ is the electron density and $G(\varrho(\vec{r}))$ is a universal function of $\varrho(\vec{r})$.

The expression for the energy E in Equation 291 is a minimum if the density functional $\varrho(\vec{r})$ is exact. Following Kohn and Sham, one develops (with an appropriate approximation for $\varrho(\vec{r})$) a scheme which is analogous in simplicity to Hartree's method, but contains the major part of the effect of Coulomb exchange and correlation. In the Kohn-Sham scheme the Coulomb correlation has been incorporated in the energy functional rather than in the wave function itself. Thus it offers a convenient way of including many-body effects with the essential feature that it retains the simplicity of calculations as in the Hartree method.

The Kohn-Sham scheme does not include the spins of the electrons explicitly. For investigating the properties of systems where the spin effects are crucial, the spin-dependent version[131-135] of the Kohn-Sham scheme may be employed. The generalizations of the Kohn-Sham scheme lead to the spin-density functional (SDF) formalism. In this SDF formalism, for a nonhomogeneous system of electrons in an external potential $V(\vec{r})$ one first obtains an equation for the ground state energy E analogous to Equation 291 and then, following Kohn and Sham, one derives a system of Hartree-like equations

$$\left[-\frac{\hbar^2}{2m}\nabla^2 + V(\vec{r}) + V_H(\vec{r}) + V_s^{xc}(\vec{r})\right]\Psi_{\nu s}(\vec{r}) = \epsilon_{\nu s}\Psi_{\nu s}(\vec{r}) \tag{292}$$

with

$$\rho_s(\vec{r}) = \sum_{\nu} |\Psi_{\nu s}(\vec{r})|^2 \tag{293}$$

$$V_H(\vec{r}) = \frac{e^2}{4\pi\epsilon_0}\int \frac{\rho(\vec{r}')}{|\vec{r}-\vec{r}'|}\,d^3r' \tag{294}$$

The summation in Equation 293 is to be taken over only those states ν which are occupied by the electrons. $V(\vec{r})$, in general, is a spin-dependent external potential which accounts for the interaction between the electrons and the nucleus and can allow spin-dependent interactions, such as the interaction between the electron spins and the applied magnetic field (if present). s is a spin index (s = + or − for "spin up" and "spin down" states, respectively). $\varrho_s(\vec{r})$ are the two components $\varrho_+(\vec{r})$ and $\varrho_-(\vec{r})$ of the total spin, such that the total density is

$$\rho(\vec{r}) = \rho_+(\vec{r}) + \rho_-(\vec{r}) \tag{295}$$

V_s^{xc} is the exchange correlation potential which, in the local spin-density approximation (LSD), is given by

$$V_s^{xc} = \frac{e^2}{4\pi\epsilon_0}\,\frac{\partial}{\partial\rho_s(\vec{r})}\left[\rho(\vec{r})\,\epsilon^{xc}(\rho_+(\vec{r}),\rho_-(\vec{r}))\right] \tag{296}$$

where ε^{xc} $(\varrho_+(\vec{r}), \varrho_-(\vec{r}))$ is the exchange correlation energy per particle for a homogeneous system of electrons. Numerical values of ε^{xc} $(\varrho_+(r), \varrho_-(r))$ have been published in the literature. However it is convenient, without loss of accuracy, to use the interpolation formula of Gunnarsson and Lundquist[134] which can be expressed in terms of the electron parameters r_s and ζ defined by

$$\frac{4}{3}\pi r_s^3 a_o^3 = \frac{1}{\rho(\vec{r})} \qquad (a_o = \text{Bohr radius}) \tag{297}$$

$$\zeta = \frac{\rho_+(\vec{r}) - \rho_-(\vec{r})}{\rho(\vec{r})}$$

as

$$\epsilon^{xc}(r_s,\zeta) = \epsilon_P^{xc} + (\epsilon_F^{xc} - \epsilon_P^{xc})\, f(\zeta) \tag{298}$$

where

$$f(\zeta) = \frac{[(1+\zeta)^{4/3} + (1-\zeta)^{4/3} - 2]}{(2^{4/3} - 1)}$$

$$\epsilon_i^{xc} = \epsilon_i^{x} - C_i\left[\left(1+x_i^3\right)\ell n\left(1+\frac{1}{x_i}\right) + \frac{1}{2}x_i - x_i^2 - 1/3\right]$$

$(i = P, F)$

with

$$x_i = \frac{r_s}{r_i}$$

$$\epsilon_P^{x} = -\frac{3}{2\pi\alpha r_s}$$

$$\alpha = \left(\frac{4}{9\pi}\right)^{1/3}$$

$$\epsilon_F^{x} = 2^{1/3}\epsilon_P^{x}$$

$$C_P = 0.0666$$

$$C_F = 0.0406$$

$$r_P = 11.4$$

$$r_F = 15.9$$

and

$$V_s^{xc} = -\frac{2}{\pi\alpha r_s}\left(\beta \pm \frac{1}{3}\frac{\delta\zeta}{1 \pm \gamma\zeta}\right) \tag{299}$$

$$\beta = 1 + 0.0545\, r_s \ln\left(1 + \frac{11.4}{r_s}\right)$$

$$\delta = 1 - 0.036\, r_s - \frac{1.36\, r_s}{1 + 10\, r_s}$$

$$\gamma = 0.297$$

The total energy of the electronic system is given by

$$E = -\frac{\hbar^2}{2m}\sum_{\nu,s}^{occ}\int \Psi^*_{\nu s}(\vec{r})\, \nabla^2\, \Psi_{\nu s}(\vec{r}) d^3r + \sum_s \int V_s(\vec{r})\, \rho_s(\vec{r})\, d^3r$$
$$+ \tfrac{1}{2}\int \rho(\vec{r})\, V_H(\vec{r})\, d^3r + \int \rho(\vec{r})\, \epsilon^{xc}(\rho_+(\vec{r}), (\rho_-(\vec{r}))d^3r \tag{300}$$

Equations 292 to 294 should be solved self-consistently. The total energy is obtained from Equation 300.

The energy parameters $\varepsilon_{\nu s}$ should not be treated as excitation energies.[135] The excitation energies can, however, be obtained by calculating the total energies of the excited states (particularly the lowest state with any given symmetry) using Equations 292 to 300 in the same way as for the ground state[134,135] and obtaining the difference between the total energies of the two states.

In the LSD approximation[131-135] the SDF theory, though formally correct for the slow and weak spatial variation of the density, has been proved by Gunnarsson and co-workers[134,135] to give a good account of the exchange and correlation energy even in the case of nonhomogeneous distribution of electron density, because only the spherical average of the exchange-correlation hole affects the energy and because it satisfies the sum rule giving single unit charge on the hole.

Gunnarsson and Lundquist[134] have proved the usefulness of the SDF theory by performing calculations on atoms, molecules, and solids. The calculated ionization potentials, affinities, and excitation energies for atoms show that the valence electrons can accurately be described; an error of less than 0.5 eV was found in the ionization energy. A broad range of interesting applications of the SDF formalism can be found in the literature.[136-143]

The importance of the SDF formalism appears to come from the realization that the correlation effect (which is neglected in the Hartree-Fock theory) plays an important part when one calculates the energy differences between two electronic configurations,[144] as for example, in obtaining excitation energies.

Some articles[144-148] on the density functional approach have come to the author's attention, which reveal the recent interest in this field.

5. Pseudopotential Theory

From the foregoing treatment it is obvious that as the number of electrons in an atom increases, the effort required in computing the eigenfunctions and eigenvalues increases tremendously. The quantum mechanical calculations for molecules and solids which make use of the atomic eigenfunctions and eigenvalues become even more tedious and complicated owing to the involvement of a large number of electrons. Also, there are many properties of atoms (and also molecules and solids) which can be interpreted in terms of the valence electrons. In such cases, the method of pseudopotentials (or effective potentials) has been found to be a useful tool. The pseudopotential method is a technique which reduces the quantum mechanical problem of $N_c + N_v$ electrons to a problem of N_v valence electrons. The N_c core electrons are indirectly taken into account in terms of a new potential, a pseudopotential, in the Hamiltonian for the N_v valence electrons. On account of this simplification the pseudopotential method has been developing rather rapidly and has already found wide applications.

Hellman[149] seems to have first introduced the concept of the pseudopotential in 1935 and, independently, Gombas[150] proposed it in 1941. In fact, they employed the model potentials which comprised a functional form of the potential with unknown parameters that were determined by the known experimental data. Fock, Vesselow, and Petrashen[151] formulated the pseudopotential problem by treating the valence electrons as a system under the influence of the remaining electrons. Also, the understanding of the statistical nature of the Thomas-Fermi model of the atom has helped in the conception of a reasonable form of the pseudopotential. Several authors[152-163] proposed different functional forms of the potential in a variety of problems. Phillips and Kleinmann[164] have developed the pseudopotential formalism in the Hartree-Fock approximation, valid only for a closed-shell atomic system where the valence and core orbital functions are eigenfunctions of the same Hamiltonian. This formalism can be generalized by relaxing the restrictions involved in it and merely requiring that the core orbitals are orthogonal to the valence orbitals. An attempt in this direction has been made by Weeks et al.[165] For the application of pseudopotentials to solid state problems, attention may be drawn to the book by Harrison.[166]

Several *ab initio* effective core potentials appropriate to the valence electrons have recently been considered by Kahn et al.[167] who have also given a useful formulation of the problem. In their approach, the wave functions and eigenvalues obtained by Hartree-Fock calculations on atoms are used to derive pseudopotentials and pseudowave functions. They could obtain pseudowave functions (for any azimuthal quantum numbers) without any nodes, as required, since as a solution of the Schrödinger equation containing the pseudopotential the valence electron state of the given ℓ should correspond to the lowest energy state. Following Kahn et al.,[167] the pseudopotential may be written as

$$V_p = U_\ell + \frac{N_c}{r}\,\frac{e^2}{4\pi\epsilon_0} \tag{301}$$

where N_c is the number of core electrons. U_ℓ is the effective core potential,

$$U_\ell = \epsilon_{n\ell} - \frac{hX}{X_{n\ell}} + \frac{W_{n\ell}X_{n\ell}}{X_{n\ell}} \tag{302}$$

where

$$h = -\frac{\hbar^2}{2m}\nabla^2 - \frac{e^2}{4\pi\epsilon_0}\,\frac{Z}{r} \tag{303}$$

$\varepsilon_{n\ell}$ is energy of the valence electron in the quantum state n_ℓ given by the pseudowave function $X_{n\ell}$. $W_{n\ell}$ accounts for the interaction energy between the valence electrons,

Table 12
TABULATION OF VARIOUS PSEUDOPOTENTIALS; N_c AND N_v ARE THE NUMBER OF CORE AND VALENCE ELECTRONS, RESPECTIVELY;

$P_\ell = \sum_{m=-\ell}^{\ell} |\ell m\rangle\langle \ell m|$ and $P_c = \sum_c |\phi_c\rangle\langle\phi_c|$, ϕ_c BEING THE CORE ORBITAL FUNCTIONS; J_c AND K_c ARE THE COULOMB AND EXCHANGE TERMS INVOLVING CORE ORBITALS; ε_c AND ε_v ARE THE CORE AND VALENCE ORBITALS ENERGIES, RESPECTIVELY

V_p	Ref.
$\Sigma_c[(\epsilon_v - \epsilon_c) P_c + 2J_c - K_c] - \frac{N_c}{r}$	Phillips and Kleinman[164]
$\Sigma_c (\epsilon_v - \epsilon_c) P_c - \frac{N_c}{r} \exp[-\alpha r]$	Coffey et al.[168]
$\Sigma_i^{n_a} \frac{N_v}{r} A_i \exp[-\alpha_i r^2] + \Sigma_c B_c P_c$	Bonifacic and Huzinga[169]
$\sum_{\ell=0}^{\infty} \Sigma_i P_\ell A_{\ell,i} r^{n_{\ell,i}} \exp[-\alpha_{\ell,i} r^2]$	Kahn and Goddard[170] Melius et al.[171,172] Kahn et al.[167] Durand and Barthelat[173]

which is present if more than one valence electron is involved. Evidently U_ℓ is different for different ℓ values. The above equations have been deduced by using the Hartree-Fock approximation, taking into account the interactions with the core electrons and imposing the orthogonality condition between the valence electrons and the core electrons. The pseudo-orbitals $X_{n\ell}$ are the solutions of the Schrödinger equation containing the pseudopotential V_p. Explicitly,

$$\left[-\frac{\hbar^2}{2m}\nabla^2 - \frac{ZN_v e^2}{4\pi\epsilon_o r} + V_p\right] X_{n\ell} = \epsilon_{n\ell} X_{n\ell} \qquad (304)$$

Examples of various *ab initio* pseudopotentials have been listed in Table 12 along with relevant references.[164,168-173] These pseudopotentials have been discussed by Tiopol et al.[174] in detail. For multivalence-electron atoms it is generally difficult to include a proper treatment of the valence-valence Coulomb and exchange interactions.[172] On the other hand in semiempirical pseudopotentials one can consider, using the appropriate parameters, the correlation between the valence and core electrons; in *ab initio* methods such correlation effects are usually ignored.

Kahn et al.[167] have used the pseudopotential approach to extract pseudopotentials for the first-row atoms and for the halogen atoms. The pseudopotentials for the third-row elements K through Zn have been developed by Tiopol et al.[174] Improvements over the original pseudopotentials for removing the large range attractive tail have recently been suggested by Hay et al.[175]

Pseudopotentials have also been introduced in the density functional and Hartree-Fock-Slater approaches.[176-180] The relationship between the full density functional and the pseudodensity functional has been considered by Harris and Jones.[176] They concluded that if the pseudopotential is carefully chosen to minimize the differences between the valence orbitals and pseudo-orbitals for the Si atomic ground state (a requirement different from that of Appelbaum and Hamann[181]) the pseudopotential approach gives good results for atomic excitation energies (and also for binding energy curves for Si_2 dimer). A method of steepest-descent has been applied to calculate the pseudo-orbitals, and the correlation effects incorporating the density functional formalism have been treated by Stoll et al.[179] A simple procedure has recently been pointed out by Hamann et al.[180] to extract norm-conserving pseudopotentials from *ab initio* atomic calculations which are found to yield the required eigenvalues and the nodeless eigenfunctions for the valence electrons. Pseudopotential methods have recently been exploited to obtain an explanation for many properties of a variety of systems.[182-184]

Though the pseudopotential method is an approximate method, it turns out to be a useful tool to study the electronic structure of complicated systems. This approach must, of necessity, introduce errors into the description of the valence electrons, and sometimes physically misleading conclusions could be drawn from the pseudopotential calculations.[178] A good and simple account has recently been given by Goodfriend[185] to establish restrictions in the use of pseudopotential results by analyzing the exact pseudopotentials for excited states of the hydrogen atom. It was inferred that the pseudopotentials could lead to correct results only for total energies, electronic energy levels, dissociation energies, and equilibrium positions of the atoms for a system. Other properties, on the other hand, need not be reliable if obtained by pseudopotential calculations.

IV. CHARACTERISTIC X-RAYS AND OPTICAL TRANSITIONS

A. Introduction

X-ray line spectra are excitingly important from theoretical as well as from practical points of view. Theoretically, they provide information as to the energy levels of the inner subshells of atoms; practically, they have found many scientific and technological applications.

X-ray line spectra are produced in an atom which is in a highly excited state and returns to its ground state by emitting a set of high energy (i.e., high frequency) photons. These photons constitute the X-ray lines which are characteristic of the atom involved in excitation and de-excitation. Figure 22 illustrates an X-ray tube in which the electrons are emitted from a heated cathode and acquire high kinetic energy (about 10^4 eV) because of the potential gradient between the anode and the cathode. The high energy electrons strike the atoms of the anode and knock off electrons from the inner subshells of the atoms. Evidently, the atoms are in the excited state and the electrons from the outer subshells then drop into the vacancies in the inner shells, thereby emitting a series of spectral lines of energy equivalent to the energy lost by the electrons in various jumps from one level to the other. Anodes consisting of heavy elements emit X-rays more strongly relative to the anodes made of light elements. Also, the square root of the frequencies of the emitted radiations have been found by Moseley[186] to increase precisely with the atomic number of the atom constituting the anode.

There are, in addition, excitations and de-excitations of atoms present in the gas discharge tube corresponding to the low energy levels of the atomic electrons. In fact, the de-excitation of the atoms corresponding to the low energies constitutes the optical line spectra. Precisely, in this process first an electron from the outer subshell of an atom is excited to its low lying energy excited states; then the excited electron returns to its ground state, emitting photons of energy equal to the energy lost by the electron in going from the upper level to the lower level. These photons fall in the optical frequency range and constitute an important research tool in many branches of phys-

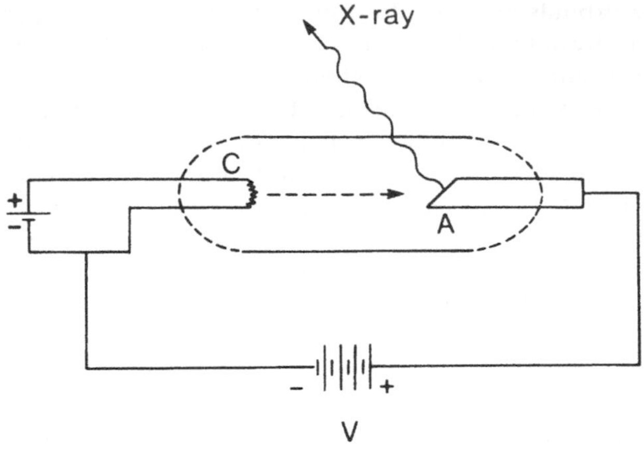

FIGURE 22. Production of X-rays. C is the cathode and A is the anode. Electrons are accelerated by the potential difference between the cathode and the anode and strike the anode to emit X-rays.

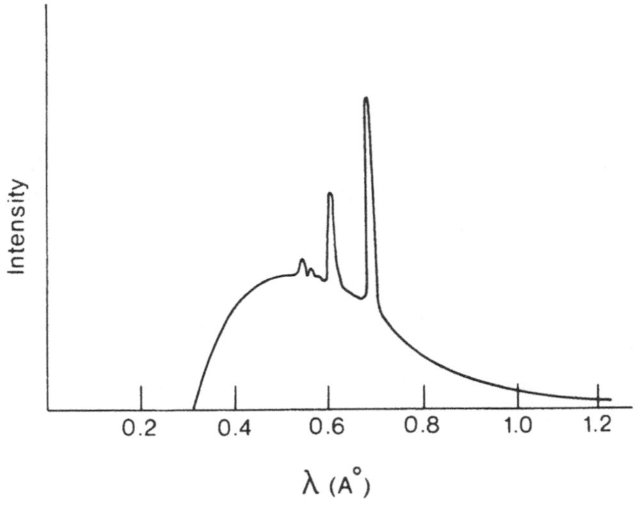

FIGURE 23. Illustration of a typical X-ray spectrum. On the continuous spectrum (the "white" radiation) is superimposed a number of bright lines characteristic of the anode material.

ics. For instance, in astronomy, optical photons have played a great role in revealing the composition and physical environment of stars. The characteristic emission spectra of stars have led to the identification of various elements in the stars and the intensity of the observed lines has provided information about the temperatures of the stars. The Doppler shift of the spectral lines has given a measure of the velocity of the star and the Zeeman effect has furnished the magnitude of the magnetic field present on the star.

The complete spectrum of the emitted radiations from the X-ray tube contains not only the discrete line spectrum but also the continuous spectrum — the "white" radiation (see Figure 23). The continuous spectrum is produced by the bremsstrahlung processes, in which the electrons from the beam in the tube are decelerated in scattering from the atoms of the anode. Though the line spectrum is characteristic of the atoms contained in the anode, the shape of the bremsstrahlung continuum depends on the energy of the electron beam.

Historically, X-rays were discovered quite accidentally by Roentgen in 1895 by observation that these rays can pass through a thick (about 10 mm) optically opaque material and produce fluorescence in certain salts of metals. Numerous scientists[186-189] have investigated the properties of X-rays. It is found that anodes (in the X-ray tube) composed of heavy elements produce X-rays more strongly than do the light elements. Also, X-rays can be diffracted by crystals, which proves that crystals are composed of regular arrangements of atoms in space. The diffraction properties of X-rays by crystals have been found to be extremely useful (by analyzing the X-ray beam in terms of its monochromatic components) for studying the atomic structure of crystals and macromolecules. It is because of X-rays that our knowledge of the basic properties of matter has tremendously increased.

In the following the characteristic X-ray spectra of atoms will be analyzed in terms of quantum mechanics. The spectroscopic notations and the selection rules will be introduced. In addition, some details of the optical excitations of various atoms will be given and the effect of the magnetic field on the spectra (Zeeman effect) will be briefly considered.

B. Characteristic X-Rays of Atoms: Selection Rules and Transition Probabilities

As mentioned earlier, the anode in the X-ray tube (see Figure 22) emits well-defined frequencies of electromagnetic waves — the characteristic radiation of the anode atoms. These radiations can be interpreted in terms of the shell structure of the atoms. We have seen that, in the shell description, an atom contains two electrons in the 1s state, two in the 2s state, six in the 2p state, and so on. Now, any one or more electrons can be excited to the higher levels of the atom, thereby creating vacancies in the lower states. The electrons fall back to the vacancies, emitting radiations which are characteristic of the energy differences of the levels of the atoms. Clearly, the use of the single-particle energy levels of the electrons to explain the emitted radiations forms an approximate procedure, since the excited states and the ground state pertain to the properties of the whole atom which must be properly accounted for in a realistic calculation. It must be remembered also that the Pauli exclusion principle is followed in the excitation and de-excitation process of the electrons from one level to the other.

The notations adopted in X-ray emissions are somewhat different than the usual notations of optical spectroscopy even though the fundamental mechanisms involved in both are precisely the same. The notations used for the values of the principal quantum number n are the capital letters K, L, M . . . for $n = 1, 2, 3 \ldots$, in accordance with the scheme introduced by Barka as shown in Table 13. The lowest shell is K which corresponds to $n = 1$. The electrons occupying the K shell are referred to as K electrons. Those occupying the L shell are the L electrons, and so on. Evidently there is a maximum of 2 K electrons, 8 L electrons, 18 M electrons, etc.

The substates of a given n are further identified by means of the Roman subscripts I, II, III, etc., in order of increasing energy (see Figure 24). The K shell has one level but there are three L subshells denoted by L_I, L_{II}, and L_{III} corresponding to n, ℓ, j equal to the set 2,0, ½; 2, 1, ½; and 2, 1, 3/2, respectively. Usually these notations are used to identify the state of excitation of an atom. Accordingly, L_{III} refers to an excited state of an atom in which an electron from the state $n = 1$, $\ell = 2$, $j = 3/2$ of the atom is missing. Thus, the important difference between the standard energy level diagram and the X-ray energy level diagram is that the latter is concerned with the

Table 13
X-RAY SPECTROSCOPIC NOTATIONS FOR VARIOUS VALUES OF THE PRINCIPAL QUANTUM NUMBER n

n	1	2	3	4	5	———
X-ray spectroscopic notations	K	L	M	N	O	———

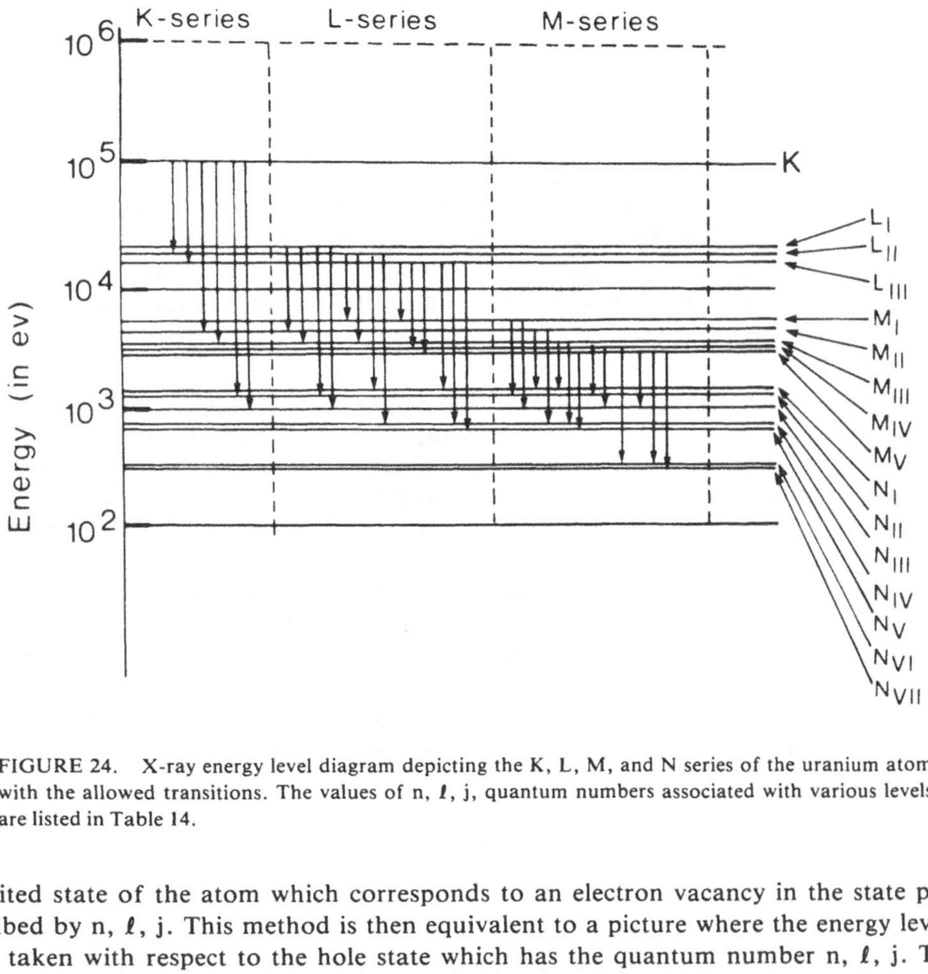

FIGURE 24. X-ray energy level diagram depicting the K, L, M, and N series of the uranium atom with the allowed transitions. The values of n, ℓ, j, quantum numbers associated with various levels are listed in Table 14.

excited state of the atom which corresponds to an electron vacancy in the state prescribed by n, ℓ, j. This method is then equivalent to a picture where the energy levels are taken with respect to the hole state which has the quantum number n, ℓ, j. The energy of a hole is positive since it represents the absence of an electron from the negative energy level. Thus all energy levels in the X-ray diagram are positive and appear to be inverted from the standard optical level diagram. Accordingly, in the X-ray energy level diagram, the K shell has the highest energy; next to it and lower in energy is the L shell, and so on. As an illustration, we depict the X-ray energy level diagram for the uranium atom and the various allowed transitions showing the K, L, M, etc. series in Figure 25. The n, ℓ, j values (as expected from the Hartree theory) associated with various energy levels of Figure 24 have been listed in Table 14. One notes that the energies of the X-ray levels decrease with an increase of n values, consistent with the hole picture as explained above. According to the inversion of the energy levels in this picture, the level corresponding to $j = \ell + \frac{1}{2}$ has a lower energy than the one with $j = \ell - \frac{1}{2}$ (see Figure 24 and Table 14). In Table 15 we have listed

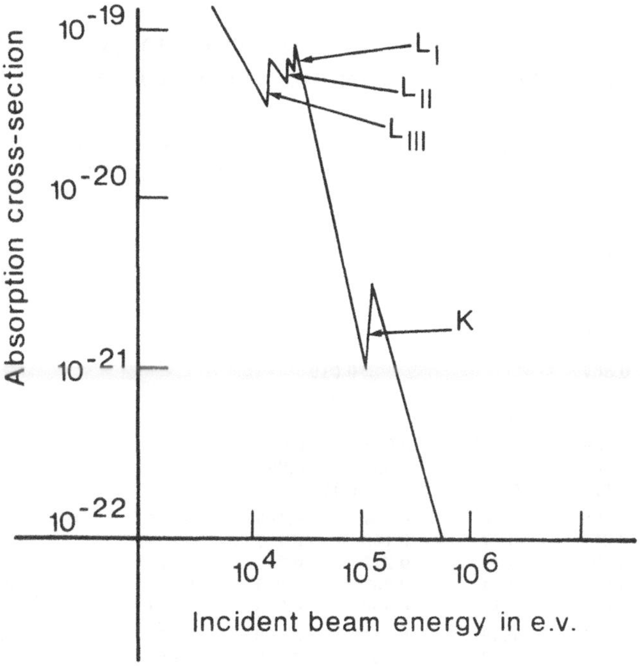

FIGURE 25. K and L absorption edges of Pb.

Table 14

n, ℓ, j VALUES ASSOCIATED WITH
VARIOUS COMPONENTS OF K, L, M,
AND N SERIES

Symbols	n	ℓ	j	Terms (symbols)
K	0	0	½	$2_{S_{1/2}}$
L$_I$	2	0	½	$2_{S_{1/2}}$
L$_{II}$	2	1	½	$2_{P_{if2}}$
L$_{III}$	2	1	3/2	$2_{P_{3/2}}$
M$_I$	3	0	½	$2_{S_{1/2}}$
M$_{II}$	3	1	½	$2_{P_{1/2}}$
M$_{III}$	3	1	3/2	$2_{P_{3/2}}$
M$_{IV}$	3	2	3/2	$2_{D_{1/2}}$
M$_V$	3	2	5/2	$2_{D_{3/2}}$
N$_I$	4	0	½	$2_{S_{1/2}}$
N$_{II}$	4	1	½	$2_{P_{1/2}}$
N$_{III}$	4	1	3/2	$2_{F_{3/2}}$
N$_{IV}$	4	2	3/2	$2_{D_{3/2}}$
N$_V$	4	2	5/2	$2_{D_{5/2}}$
N$_{VI}$	4	3	5/2	$2_{F_{5/2}}$
N$_{VII}$	4	3	7/2	$2_{F_{7/2}}$

the energies of various levels (actually, the binding energies of electrons in these levels) of all atoms involved in X-ray emissions.

From the foregoing treatment it is clear that X-rays can be pictured as produced by the jumping of a hole from one state to the other. Created with each jump is a photon

Table 15
BINDING ENERGY OF ELECTRONS IN UNITS OF 10^3 eV FOR DIFFERENT SHELLS IN ATOMS USEFUL FOR DERIVING X-RAY ENERGIES[232]

	K	L_I	L_{II}	L_{III}	M_I	M_{II}	M_{III}	M_{IV-V}
3 Li	0.055							
4 Be	0.112							
5 B	0.188							
6 C	0.284							
7 N	0.400							
8 O	0.532							
9 F	0.68							
10 Ne	0.867		0.019					
11 Na	1.061		0.031					
12 Mg	1.303	0.063	0.049					
13 Al	1.559	0.117	0.073	0.072				
14 Si	1.838	0.12	0.099	0.098				
15 P	2.142	0.15	0.128					
16 S	2.470	0.19	0.164	0.163	0.015			
17 Cl	2.819	0.24	0.199	0.197	0.014			
18 Ar	3.203	0.29	0.247	0.245	0.027	0.012		
19 K	3.608	0.34	0.297	0.294	0.034	0.018		
20 Ca	4.038	0.40	0.350	0.346	0.044	0.025		
21 Sc	4.497	0.46	0.411	0.406	0.058	0.036		
22 Ti	4.965	0.53	0.459	0.453	0.058	0.033		
23 V	5.464	0.60	0.519	0.511	0.065	0.036		
24 Cr	5.989	0.680	0.583	0.574	0.073	0.042		
25 Mn	6.537	0.752	0.650	0.639	0.082	0.047		
26 Fe	7.111	0.841	0.720	0.707	0.093	0.053		
27 Co	7.708	0.927	0.793	0.778	0.100	0.059		
28 Ni	8.331	1.008	0.870	0.853	0.110	0.067		
29 Cu	8.979	1.100	0.951	0.931	0.120	0.077	0.074	
30 Zn	9.661	1.197	1.045	1.022	0.140	0.089		0.010
31 Ga	10.369	1.30	1.144	1.117	0.160	0.108	0.104	0.019
32 Ge	11.104	1.42	1.248	1.217	0.180	0.126	0.122	0.029
33 As	11.864	1.529	1.356	1.320	0.200	0.144	0.138	0.038
34 Se	12.653	1.66	1.472	1.431	0.227	0.164	0.157	0.052
35 Br	13.476	1.79	1.598	1.552	0.259	0.192	0.185	0.072
36 Kr	14.324	1.904	1.726	1.675	0.291	0.220	0.211	0.087

	K	L_I	L_{II}	L_{III}	M_I	M_{II}	M_{III}	M_{IV}	M_V
37 Rb	15.201	2.067	1.866	1.806	0.324	0.250	0.240	0.114	0.112
38 Sr	16.107	2.217	2.008	1.941	0.359	0.281	0.270	0.137	0.134
39 Y	17.038	2.372	2.154	2.079	0.394	0.312	0.300	0.159	0.157
40 Zr	17.996	2.529	2.305	2.220	0.429	0.342	0.328	0.181	0.178
41 Nb	18.989	2.701	2.468	2.373	0.472	0.383	0.366	0.211	0.208
42 Mo	20.003	2.867	2.628	2.523	0.506	0.412	0.394	0.234	0.230
43 Tc	21.05	3.04	2.80	2.68	0.55	0.45	0.43	0.26	0.25
44 Ru	22.118	3.225	2.966	2.837	0.585	0.484	0.460	0.283	0.279
45 Rh	23.218	3.411	3.145	3.002	0.625	0.520	0.495	0.310	0.305
46 Pd	24.349	3.603	3.330	3.172	0.669	0.557	0.530	0.339	0.334
47 Ag	25.515	3.807	3.525	3.353	0.719	0.603	0.572	0.374	0.368
48 Cd	26.711	4.019	3.728	3.539	0.771	0.652	0.617	0.411	0.405
49 In	27.938	4.237	3.937	3.730	0.824	0.701	0.663	0.450	0.442
50 Sn	29.201	4.465	4.156	3.929	0.883	0.756	0.714	0.493	0.484
51 Sb	30.492	4.699	4.381	4.133	0.944	0.812	0.766	0.537	0.487
52 Te	31.815	4.939	4.613	4.342	1.006	0.869	0.818	0.583	0.572
53 I	33.171	5.190	4.854	4.559	1.074	0.932	0.876	0.633	0.621
54 Xe	34.588	5.453	5.101	4.782	1.15	1.00	0.941	0.67	0.67
55 Cs	35.984	5.721	5.359	5.011	1.216	1.071	1.003	0.738	0.724
56 Ba	37.441	5.994	5.624	5.247	1.292	1.141	1.066	0.795	0.780
57 La	38.931	6.270	5.896	5.490	1.368	1.207	1.126	0.853	0.838
58 Ce	40.448	6.554	6.170	5.729	1.440	1.276	1.188	0.906	0.889
59 Pr	41.996	6.839	6.446	5.969	1.514	1.340	1.246	0.955	0.935
60 Nd	43.571	7.128	6.723	6.209	1.576	1.405	1.298	1.001	0.978

Table 15 (continued)
BINDING ENERGY OF ELECTRONS IN UNITS OF 10^3 eV FOR DIFFERENT SHELLS IN ATOMS USEFUL FOR DERIVING X-RAY ENERGIES[232]

	K	L_I	L_{II}	L_{III}	M_I	M_{II}	M_{III}	M_{IV}	M_V
61 Pm	45.19	7.42	7.02	6.46	1.65	1.47	1.36	1.05	1.03
62 Sm	46.843	7.739	7.313	6.717	1.723	1.542	1.420	1.108	1.080
63 Eu	48.520	8.055	7.620	6.980	1.803	1.615	1.483	1.164	1.134
64 Gd	50.221	8.379	7.930	7.243	1.881	1.691	1.546	1.217	1.186
65 Tb	51.990	8.713	8.252	7.515	1.968	1.772	1.617	1.277	1.242
66 Dy	53.777	9.044	8.580	7.790	2.045	1.857	1.674	1.332	1.294
67 Ho	55.601	9.395	8.912	8.067	2.124	1.924	1.743	1.387	1.347
68 Er	57.465	9.754	9.261	8.357	2.204	2.009	1.680	1.451	1.407
69 Tm	59.383	10.120	9.617	8.649	2.307	2.094	1.889	1.516	1.471
70 Yb	61.313	10.488	9.979	8.944	2.398	2.174	1.951	1.576	1.529
71 Lu	63.314	10.869	10.347	9.242	2.489	2.262	2.022	1.637	1.587
72 Hf	65.323	11.266	10.736	9.558	2.597	2.362	2.104	1.713	1.658
73 Ta	67.411	11.678	11.132	9.878	2.704	2.464	2.190	1.789	1.731
74 W	69.519	12.092	11.537	10.200	2.812	2.568	2.275	1.865	1.802
75 Re	71.673	12.524	11.957	10.533	2.929	2.677	2.364	1.947	1.880
76 Os	73.872	12.967	12.385	10.871	3.048	2.791	2.456	2.029	1.959
77 Ir	76.109	13.415	12.821	11.213	3.168	2.904	2.548	2.113	2.038
78 Pt	78.392	13.875	13.270	11.561	3.294	3.020	2.643	2.199	2.118
79 Au	80.726	14.355	13.735	11.921	3.428	3.151	2.745	2.294	2.208
80 Hg	83.119	14.843	14.214	12.287	3.564	3.282	2.848	2.389	2.298
81 Tl	85.531	15.349	14.699	12.659	3.707	3.419	2.959	2.487	2.391
82 Pb	88.015	15.873	15.210	13.046	3.862	3.568	3.079	2.596	2.495
83 Bi	90.536	16.396	15.719	13.426	4.007	3.705	3.187	2.695	2.588
84 Po	93.11	16.94	16.24	13.82	4.16	3.85	3.30	2.80	2.69
85 At	95.74	17.49	16.79	14.22	4.32	4.01	3.42	2.91	2.79
86 Em	98.41	18.06	17.34	14.62	4.49	4.16	3.54	3.02	2.89
87 Fr	101.14	18.64	17.90	15.03	4.65	4.32	3.67	3.13	3.00
88 Ra	103.93	19.236	18.484	15.445	4.822	4.489	3.792	3.249	3.104
89 Ac	106.76	19.85	19.08	15.87	5.00	4.65	3.92	3.37	3.21
90 Th	109.648	20.463	19.691	16.299	5.182	4.822	4.039	3.489	3.331
91 Pa	112.60	21.105	20.314	16.374	5.367	5.002	4.175	3.612	3.441
92 U	115.610	21.756	20.946	17.168	5.549	5.182	4.303	3.726	3.551
93 Np	118.66	22.41	21.59	17.61	5.74	5.36	4.43	3.85	3.66
94 Pu	121.77	23.10	22.25	18.06	5.93	5.56	4.56	3.98	3.78
95 Am	124.94	23.80	22.94	18.52	6.14	5.75	4.70	4.11	3.90
96 Cm	128.16	24.50	23.63	18.99	6.35	5.95	4.84	4.24	4.02
97 Bk	131.45	25.23	24.34	19.47	6.56	6.16	4.99	4.38	4.15
98 Cf	134.79	25.98	25.07	19.95	6.78	6.37	5.13	4.51	4.28
99 Es	138.19	26.74	25.82	20.44	6.99	6.58	5.28	4.65	4.41
100 Fm	141.66	27.51	26.57	20.93	7.22	6.81	5.42	4.80	4.54

	N_I	N_{II}	N_{III}	N_{IV}	N_V	O_I	O_{II}	O_{III}	O_{IV-V}
37 Rb	0.031	0.017							
38 Sr	0.039	0.021							
39 Y	0.045	0.025							
40 Zr	0.050	0.026							
41 Nb	0.061	0.038							
42 Mo	0.066	0.037							
43 Tc	0.07	0.04							
44 Ru	0.074	0.043							
45 Rh	0.079	0.047							
46 Pd	0.085	0.050							
47 Ag	0.096	0.058							
48 Cd	0.108	0.068		0.011	0.010				
49 In	0.121	0.076		0.016	0.016				
50 Sn	0.137	0.088		0.025	0.024				
51 Sb	0.152	0.099		0.033	0.032	0.006			
52 Te	0.169	0.110		0.042	0.040	0.012			
53 I	0.188	0.124		0.053	0.051	0.017			
54 Xe	0.21	0.139		0.07	0.06	0.02			
55 Cs	0.230	0.178	0.167	0.078	0.075	0.022	0.018		

Table 15 (continued)
BINDING ENERGY OF ELECTRONS IN UNITS OF 10^3 eV FOR DIFFERENT SHELLS IN ATOMS USEFUL FOR DERIVING X-RAY ENERGIES[232]

	N_I	N_{II}	N_{III}	N_{IV}	N_V	O_I	O_{II}	O_{III}	O_{IV-V}
56 Ba	0.253	0.196	0.184	0.092	0.090	0.039	0.019		
57 La	0.276	0.210	0.195	0.107	0.106	0.039	0.017		
58 Ce	0.295	0.228	0.212	0.118	0.116	0.044	0.026		
59 Pr	0.309	0.240	0.222	0.123	0.118	0.042	0.023		
60 Nd	0.316	0.243	0.225	0.120	0.118	0.039	0.020		
61 Pm	0.33	0.25	0.24	0.13	0.12	0.04	0.02		
62 Sm	0.345	0.267	0.248	0.133	0.129	0.038	0.022		
63 Eu	0.363	0.287	0.258	0.140	0.138	0.035	0.027		
64 Gd	0.376	0.291	0.273	0.145	0.140	0.037	0.023		0.006
65 Tb	0.399	0.315	0.289	0.150	0.148	0.040	0.028		0.006
66 Dy	0.413	0.331	0.292	0.161	0.154	0.043	0.025		0.005
67 Ho	0.431	0.345	0.308	0.166	0.156	0.046	0.021		0.006
68 Er	0.449	0.368	0.323	0.178	0.168	0.048	0.031		0.007
69 Tm	0.472	0.390	0.341	0.191	0.181	0.053	0.036		0.009
70 Yb	0.487	0.398	0.344	0.198	0.185	0.055	0.03		
71 Lu	0.505	0.406	0.358	0.203	0.193	0.056	0.03		
72 Hf	0.535	0.434	0.377	0.220	0.210	0.061	0.035	0.028	
73 Ta	0.562	0.462	0.401	0.238	0.227	0.068	0.041	0.033	
74 W	0.589	0.485	0.418	0.252	0.239	0.070	0.040	0.030	
75 Re	0.623	0.515	0.442	0.271	0.258	0.081	0.043	0.032	
76 Os	0.654	0.544	0.467	0.289	0.273	0.084	0.056	0.044	
77 Ir	0.688	0.574	0.492	0.310	0.293	0.094	0.060	0.048	
78 Pt	0.732	0.604	0.513	0.329	0.310	0.099	0.061	0.048	
79 Au	0.762	0.647	0.548	0.354	0.337	0.111	0.076	0.057	0.005
80 Hg	0.805	0.681	0.580	0.384	0.362	0.125	0.087	0.065	0.014
81 Tl	0.848	0.725	0.613	0.409	0.389	0.139	0.101	0.078	0.017
82 Pb	0.903	0.777	0.658	0.445	0.424	0.158	0.122	0.097	0.031
83 Bi	0.945	0.814	0.688	0.472	0.447	0.167	0.126	0.102	0.033
84 Po	0.99	0.85	0.72	0.50	0.48	0.18	0.14	0.11	0.03
85 At	1.05	0.90	0.75	0.53	0.51	0.19	0.15	0.12	0.04
86 Em	1.10	0.95	0.79	0.56	0.54	0.21	0.17	0.13	0.05
87 Fr	1.15	1.00	0.83	0.60	0.57	0.23	0.18	0.14	0.06
88 Ra	1.209	1.057	0.879	0.636	0.603	0.254	0.200	0.152	0.068
89 Ac	1.26	1.11	0.92	0.67	0.64	0.27	0.21	0.16	0.08
90 Th	1.323	1.160	0.959	0.711	0.676	0.290	0.224	0.173	0.087
91 Pa	1.385	1.233	1.007	0.752	0.709	0.303	0.24	0.18	0.09
92 U	1.439	1.272	1.042	0.780	0.738	0.325	0.256	0.198	0.097
93 Np	1.50	1.32	1.08	0.82	0.77				
94 Pu	1.56	1.38	1.13	0.86	0.82				
95 Am	1.63	1.47	1.18	0.90	0.86				
96 Cm	1.70	1.51	1.22	0.94	0.90				
97 Bk	1.77	1.58	1.27	0.99	0.94				
98 Cf	1.84	1.65	1.32	1.03	0.98				
99 Es	1.91	1.72	1.37	1.08	1.03				
100 Fm	1.99	1.79	1.42	1.13	1.08				

which carries the excess energy. The frequency of the photon emitted is related to the energy E of the photon by $\nu = E/h$ (h = Planck's constant). The selection rules for the transition of the hole from one state to the other are the same as for the transition of an electron as discussed previously (see Section II). Explicitly, the selection rules are

$$\Delta \ell = \pm 1$$

$$\Delta j = 0, \pm 1$$

for the change in the azimuthal quantum number ℓ and the total quantum number j.

There are further standard notations for the constituent spectral lines of various K,

L, M, etc., series. If the transitions occur from the K shell to the L shell, the corresponding spectral lines are known as K_α lines; if the transitions occur from the K shell to the M shell, the lines are called K_β lines, etc. Thus there are K_α, K_β, K_γ, K_δ, etc., lines and similarly L_α, L_β, L_γ, L_δ lines, and so on.

As regards the quantitative interpretation of the X-ray spectra, it is obvious that one needs to calculate the various energy levels of the multielectron atom. First of all, Moseley interpreted the X-ray spectra on the basis of the Bohr model which was proposed just before his experimental measurements on X-rays. He established the correlation between the nuclear charge of an atom and its ordered position in the periodic table. He predicted the presence of elements with atomic numbers 43, 61, 72, and 75, which were subsequently discovered. As for accurate quantitative investigation of the X-ray energy levels, the theories elaborated in Section III are helpful. The details of the calculations also have already been elaborated there.

The intensity of the X-rays emitted is different for different transitions and varies with atoms. We use the following symbols (Siegbahn's notations) for the transitions $K \rightarrow Y$:

$$K_{\alpha_1} = K - L_{III}$$

$$K_{\alpha_2} = K - L_{II}$$

$$K_{\beta_1} = K - M_{III}$$

$$K_{\beta_2} = K - N_{III}$$

$$K_{\beta_3} = K - M_{II}$$

$$K_{\beta_4} = K - N_{II}$$

$$K_{\beta_5} = K - M_{IV}$$

Usually the lines $K\beta'_1$ and $K\beta'_2$ are observed in the spectra, which are resolved into $K\beta_1$, $K\beta_3$, $K\beta_5$, and $K\beta_2$, $K\beta_4$ components, respectively, under high resolution. In Table 16 we give the intensity of $K\alpha_2$, $K\beta'_1$, and $K\beta'_2$ lines per 10^3 emitted $K\alpha_1$ quanta as has been derived by Wapstra[190] from experimental data of Williams[191] ($24 \leqslant Z \leqslant 52$), Meyers[192] ($23 \leqslant Z \leqslant 49$), and Beckman[193,194] ($73 \leqslant Z \leqslant 92$).

A process opposite to emission is the absorption of photons. The probability measurements for the Pb atom have been shown in Figure 25 in which the cross section for the absorption of X-ray photons as a function of the energy of the incident photon beam in a photoelectric effect has been plotted. The discontinuity in the curve in Figure 25 near 10^5 eV is known as the K absorption edge, whereas the discontinuities near 10^4 eV are the L absorption edges. The K absorption edge corresponds to the energy required for the production of a hole in the K shell. There are three L absorption edges since the L level has been split into three levels due to spin-orbit and other relativistic effects. Clearly, the L absorption discontinuities depict the energies required to produce holes in L_I, L_{II} and L_{III} levels. Similar absorption spectra are expected for other atoms which reveal various absorption edges. From the data of Allen[187] it is possible to exhibit[188] various absorption edges in several elements. The absorption edges were first discovered by de Broglie.[189]

C. Coupling of Angular Momenta, LS Coupling, and Spectroscopic Symbols

We have seen in Section II that in the Hartree picture an electron can be thought of

Table 16
RELATIVE K X-RAY INTENSITIES IN TERMS OF INTENSITY COMPONENTS

Atomic number	X-ray intensity components		
	α_2/α_1	β'_1/β_1	β'_2/β_1
20	0.520	0.195	
22	0.514	0.180	
24	0.509	0.169	0.005
26	0.504	0.162	0.003
28	0.500	0.168	0.002
30	0.499	0.180	0.003
32	0.498	0.192	0.006
34	0.498	0.202	0.013
36	0.499	0.212	0.021
38	0.500	0.220	0.028
40	0.500	0.228	0.034
42	0.501	0.236	0.040
44	0.503	0.243	0.044
46	0.505	0.250	0.048
48	0.507	0.257	0.051
50	0.510	0.264	0.054
52	0.513	0.270	0.056
54	0.516	0.276	0.058
56	0.519	0.282	0.060
58	0.521	0.287	0.062
60	0.524	0.294	0.064
62	0.526	0.300	0.066
64	0.529	0.305	0.068
66	0.532	0.310	0.070
68	0.535	0.316	0.072
70	0.537	0.322	0.075
72	0.540	0.327	0.078
74	0.542	0.333	0.081
76	0.545	0.338	0.085
78	0.547	0.344	0.090
80	0.550	0.349	0.096
82	0.553	0.355	0.102
84	0.555	0.360	0.108
86	0.558	0.366	0.115
88	0.560	0.371	0.122
90	0.563	0.377	0.130
92	0.565	0.382	0.138
94	0.567	0.388	0.146
96	0.570	0.393	0.154
98	0.573	0.399	0.162
100	0.576	0.404	0.170

From Wapstra, A., Nijgh, G., and Van Lieushout, R., *Nuclear Spectroscopy Tables*, North-Holland, Amsterdam, 1959. With permission.

as moving independently in the average field of the nucleus and other electrons in the atom, and its state can be described by its quantum numbers n and ℓ. The Hartree approximation takes into account the dominant interaction of the electron in question with the nucleus and other electrons. In order to explain the optical spectra one needs to include the further, weaker, interactions which are (1) the residual Coulomb interaction which corrects the Hartree potential to account for the remaining weaker part of the interaction arising from the nucleus and other electrons, (2) the spin-orbit interaction which couples the spin angular momentum of an electron with its orbital angular momentum, and (3) the interaction of the spin of one electron with the spin of another electron arising because of the magnetic interactions between the spin magnetic moments. The last effect is very small and is usually neglected.

We have already acquainted ourselves with the spin-orbit interaction and the coupling of the spin and orbital angular momenta in Section II.

The theories we have discussed in Section III can systematically take into account the corrections over the Hartree approximation. One, of course, has to satisfy the asymmetric property of the total wave function as elaborated in Section III. For quantitative considerations in the first approximation, the expectation values of the residual Coulomb and spin-orbit interactions must be added to the energies obtained from the Hartree approximation.

Because of the residual Coulomb interaction one electron exerts torques on the other, particularly when the electron charge distributions are not spherically symme-

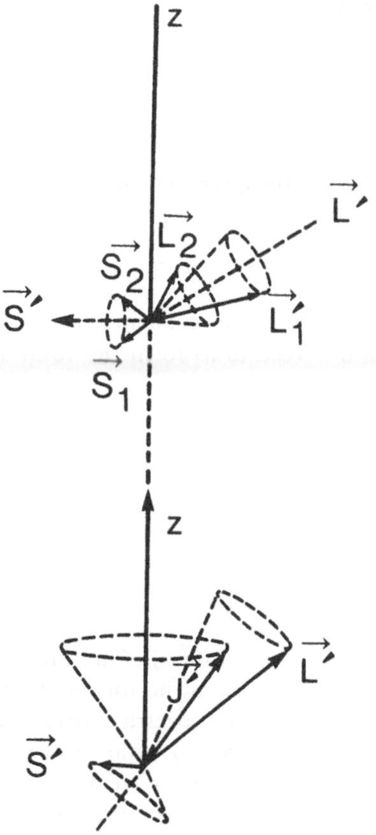

FIGURE 26. Vector representation of the coupling of spin and orbital angular momenta of two electrons for the total angular momentum $\vec{J}'$.

tric. The effect of this is that the magnitudes of the individual orbital angular momenta of the electrons do not get changed. The torques only make the individual angular momenta precess about the total orbital angular momentum vector $\vec{L}'$ in such a way that the magnitude of $\vec{L}'$ remains constant. This leads us directly to the coupling of angular momenta of electrons (see Figure 26).

The total angular momentum $\vec{L}'$ can be formed from the individual orbital angular momenta $\vec{L}_1$, $\vec{L}_2$, $\vec{L}_3$, + . . . of electrons as

$$\vec{L}' = \vec{L}_1 + \vec{L}_2 + \vec{L}_3 + \ldots \vec{L}_i + \ldots$$

with the magnitude of $\vec{L}'$ given in terms of the corresponding quantum number ℓ' by the quantization condition

$$L' = \hbar\sqrt{\ell'(\ell' + 1)} \qquad (305)$$

The coupling is accomplished in such a manner that all $\vec{L}_i$s remain constant. The individual spin angular momenta $\vec{S}_i$ of the electrons also couple (see Figure 26) to form the total spin angular momentum $\vec{S}$ as

$$\vec{S}' = \vec{S}_1 + \vec{S}_2 + \vec{S}_3 + \ldots \vec{S}_i + \ldots \tag{306}$$

with the magnitude of $\vec{S}'$ given by the quantization rule

$$S' = \hbar\sqrt{s'(s'+1)} \tag{307}$$

The weaker interaction, viz., the spin-orbit interaction, can now be taken into account by coupling the $\vec{L}$ and $\vec{S}$ vectors to form the total angular momentum $\vec{J}'$ given by (see also Figure 26)

$$\vec{J}' = \vec{L}' + \vec{S}' \tag{308}$$

with the quantization condition

$$J' = \hbar\sqrt{j'(j'+1)} \tag{309}$$

in terms of the associated quantum number j′. The method described is known as LS coupling, or Russell-Saunders coupling in honor of the two scientists who used this coupling procedure to explain the atomic spectra emitted by stars.

LS coupling is appropriate for atoms of small and intermediate Z values since in these atoms residual Coulomb interaction is strong and the spin-orbit interaction is much weaker. For atoms of large Z, however, the spin-orbit interaction is very large, which necessitates first coupling the spin and orbital angular momenta of each electron to form the total angular momenta. Accordingly.

$$\vec{J}_i = \vec{L}_i + \vec{S}_i \tag{310}$$

The next step is to form the total angular momentum $\vec{J}$ by coupling all $\vec{J}_i$s,

$$\vec{J}' = \sum_i \vec{J}_i \tag{311}$$

This procedure is known as JJ coupling and is important for atoms of high Z-values and for the behavior of protons and nucleons in nuclei where the spin-orbit interaction for each particle is very strong.

If there are two valence electrons such that the configuration of the valence electrons can be specified as 4p4d, then, in accordance with the above treatment, there are 12 terms which have been shown schematically in Figure 27.

In the LS coupling procedure the spin angular momenta $\vec{S}_i$ precess about their sum S′ the orbital angular momenta precess around their sum $\vec{L}'$, and the total spin angular momentum $\vec{S}'$ and the total orbital angular momentum L′ precess about their sum, $\vec{J}'$. Finally the total angular momentum $\vec{J}'$ precesses about the Z-axis in order to follow the angular momentum uncertainty principle with the quantization condition for the Z-component of the total angular momentum,

$$J'_z = m_{j'}\hbar \tag{312}$$

$$\left\{\begin{matrix}\text{Unperturbed}\\\text{configuration}\end{matrix}\right\} + \left\{\begin{matrix}\text{Spin-spin}\\\text{correlation}\\\text{energy}\end{matrix}\right\} + \left\{\begin{matrix}\text{Residual}\\\text{electrostatic}\\\text{interaction}\end{matrix}\right\} + \left\{\begin{matrix}\text{Spin-Orbit}\\\text{interaction}\end{matrix}\right\}$$

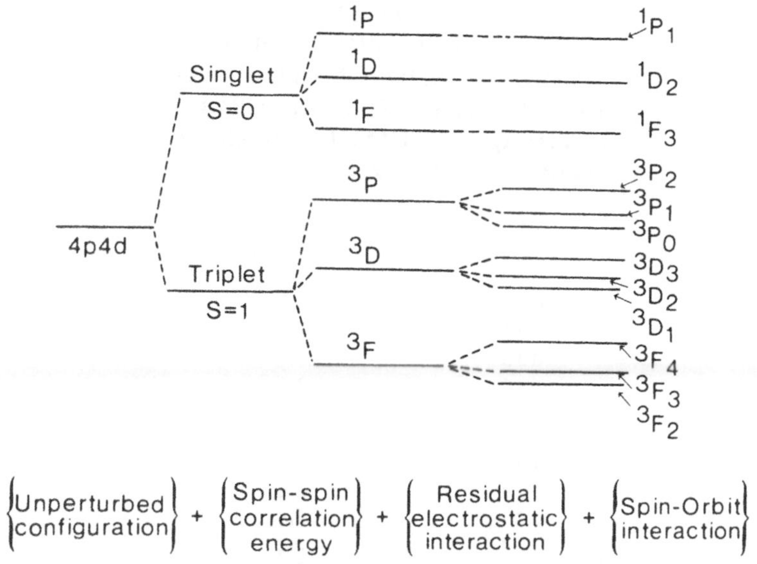

FIGURE 27.   Illustration of the formation of spectroscopic terms for the case of two electrons in the configuration 4p4d.

where

$$m_{j'} = -j', -j'+1, \ldots, j'-1, j' \tag{313}$$

Figure 26 shows schematically the coupling of angular momenta of two electrons. Obvious generalizations can be made for more than two electrons. The possible values of s', ℓ', and j' are obtained by the vector addition of the individual quantum numbers as explained in Section II. For the case of two electrons,

$$s' = |s_1 - s_2|, \quad |s_1 - s_2| + 1, \quad \ldots \quad (s_1 + s_2)$$

$$\ell' = |\ell_1 - \ell_2|, \quad |\ell_1 - \ell_2| + 1, \quad \ldots \quad (\ell_1 + \ell_2)$$

$$\text{and} \quad j' = |s - \ell|, \quad |s - \ell| + 1, \quad \ldots \quad (s + \ell) \tag{314}$$

In particular, let $\ell_1 = 2$, $s_1 = \frac{1}{2}$, and $\ell_2 = 1$, $s_2 = \frac{1}{2}$. Then one gets

$$s' = 0, 1$$

and

$$\ell' = 1, 2, 3$$

Thus,

1.   For $s' = 0$ and $\ell' = 1, 2, 3$; $j' = 1, 2, 3$.
2.   For $s' = 1$ and $\ell' = 1$; $j' = 0, 1, 2$.
3.   For $s' = 1$, $\ell' = 2$; $j' = 1, 2, 3$.
4.   For $s' = 1$, $\ell' = 3$; $j' = 2, 3, 4$.

Table 17
SPECTROSCOPIC SYMBOLS CORRESPONDING TO THE TWO-ELECTRON STATES PRESCRIBED BY THE QUANTUM NUMBERS s′ ℓ′ j′ FOR $\ell_1 = 1, \ell_2 = 2, s_1 = 1/2, s_2 = 1/2$

s′ℓ′j′	Spectroscopic symbol
0 1 1	$^1P_1$
0 2 2	$^1D_2$
0 3 3	$^1F_3$
1 1 2	$^3P_2$
1 1 1	$^3P_1$
1 1 0	$^3P_0$
1 2 3	$^3D_3$
1 2 2	$^3D_2$
1 2 1	$^3D_1$
1 3 4	$^3F_4$
1 3 3	$^3F_3$
1 3 2	$^3F_2$

Therefore there are 12 different sets of stages specified by ℓ′s′j′ in the above case; each of these sets gives rise to 2j′ + 1 values of $m_{j'}$. Every state ℓ′s′j′ can be written down in the spectroscopic notation by means of the general symbol $(2s' + 1)\ell'_{j'}$ where ℓ′ represents one of the capital letters, S, P, D, F, G, H, . . . according as ℓ′ = 0, 1, 2, 3, 4, 5 . . . In Table 17 we have listed various spectroscopic symbols for the two-electron case considered above for different sets of s′ℓ′j′. The number 2s + 1, the superscript on the capital letter, is known as the multiplicity of the spectroscopic state. The state $2s + 1_{\ell_j}$ is also referred to as the spectroscopic "term". The multiplicity of the term is read as singlet, doublet, triplet, etc., according as 2s + 1 is 1, 2, 3, etc.

As for knowing which state (or term) lies lowest in energy, it has been observed (and can be explained on physical grounds) that for an atom (with less than a half-filled shell) of small and intermediate Z, the level with largest L′ and S′ has the minimum energy. Besides this, because of the spin-orbit interaction, the energy is lowest for the state for which J′ is smallest. It must be stressed that the "terms" which do not satisfy the Pauli exclusion principle must be deleted from physical considerations.

D. Selection Rules for LS Coupling

We shall here deal with the selection rules for the transition between the energy levels of a multielectron atom. The selection rules described below correspond to electric dipole transitions and can be derived directly from the angular momentum eigenfunctions. These selection rules are similar to the ones mentioned in Section II in context with the transitions for one-electron atoms. For LS coupling the following selection rules are followed:

1. Transitions occur in such a way that only one electron "jumps" from one level to the other at a time. In other words, in transitions from one configuration to the other, only one electron changes its state.
2. The quantum number ℓ of the electron which jumps in the transition must satisfy,

$$\Delta \ell = \pm 1$$

3. Selection rules for the other quantum numbers are

$$\Delta s' = 0$$

$$\Delta \ell' = 0, \pm 1$$

$$\Delta j' = 0, \pm 1 \text{ but } j'=0 \text{ to } j'=0 \text{ is forbidden}$$

$$\Delta m_{j'} = 0, \pm 1 \text{ but } m_{j'}=0 \text{ to } m_{j'}=0 \text{ is forbidden if } \Delta j'=0 \qquad (315)$$

These rules are necessary to understand the spectral lines from the energy level diagrams for multielectron atoms.

Selection rules for jj coupling are the same as above except that the additional selection rule, i.e., $\Delta j = 0, \pm 1$ for the transition of electrons must also be followed.

E. Optical Excitations

1. Atoms with a Single Optically Active Electron

The simplest such atoms are alkali atoms in which a single electron occupies the outermost s subshell and the lower shells are completely filled. Just below the s shell lies the p subshell, where the energy is much lower than in the s subshell. The outermost s electron plays the part in low energy excitations which gives rise to optical transitions. The Hartree theory for such systems is very applicable since the net potential for the outermost (optical) electron due to the nucleus and the completely filled subshells is spherically symmetric. In fact, Hartree calculations yield very good agreement with the experimental results for alkali atoms. For illustrative purposes some of the energy levels of Na have been given in Figure 28, where the zero of the energy scale corresponds to the situation in which the outermost electron was completely ejected, leaving the atom in a singly ionized state.

Because of the spin-orbit interaction of the optically active electron the lines of the optical spectra possess a fine-structure splitting containing doublets corresponding to $j = \ell - ½$ and $j = \ell + ½$ except for $\ell = 0$. Other relativistic effects for such multielectron atoms have been found, in general, to be negligible compared with the spin-orbit interaction. The spectral lines can easily be explained by means of the energy level diagram of Figure 28 and the spin-orbit splitting by satisfying the selection rules $\Delta\ell = \pm 1$ and $\Delta j = 0, \pm 1$.

2. Atoms with Complex Spectra

If the outer subshell of an atom is partially filled with a certain number of electrons, all of these electrons are optically active and play an important part in optical excitations. These electrons can be studied in the Hartree approximation (which takes into account the dominant part of the Coulomb interaction); further correction can be made for the weaker, though important, interactions such as the residual Coulomb interaction and the spin-orbit interaction.

As we have discussed previously, the multielectron wave function must be antisymmetric with respect to the exchange of the particle. Consequently, as discussed in Section III the two electrons in the triplet state like to be more apart than when they are in the singlet state. This introduces a kind of correlation between the spin and the space coordinates. As a result, the energy of the atom is lowest when S′ is largest, where $\vec{S}'$ is the total spin vector of the electrons of the atom. We have also arrived at the conclusion that because of the residual Coulomb interactions, the orbital angular momenta of the electrons couple to give rise to the total angular momentum. In addition, spin-orbit interaction further perturbs the energy levels.

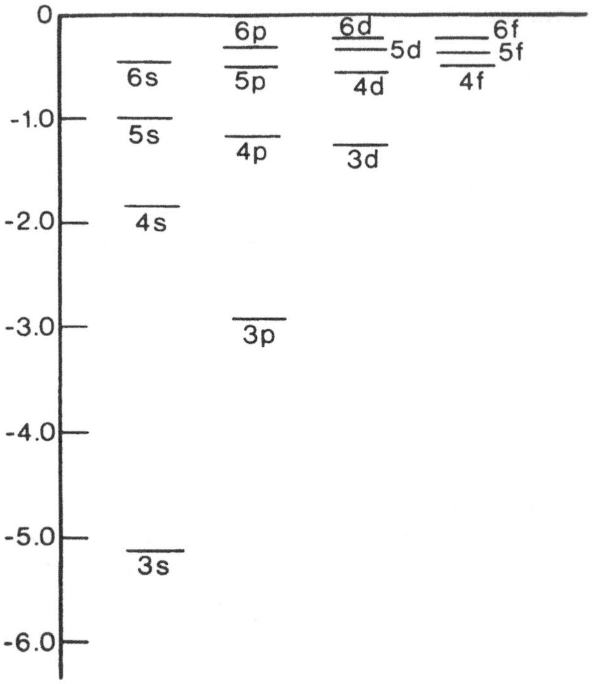

FIGURE 28. Some of the energy levels of a neutral sodium atom.

For instance we consider the energy levels of carbon, which contains six electrons in the configuration $1s^2\ 2s^2\ 2p^2$ in its ground state. In this case there are two electrons in the outermost state, namely the p state, for which the n quantum number is the same. One further recalls that the Pauli exclusion principle is not to be violated in constructing various spectroscopic terms. Figure 29 schematically illustrates the terms that one obtains for two electrons in the 2p state, which also shows that there are both singlet and triplet states. Explicitly, the $2p^2$ electrons give rise to the terms: $^1S_0$, $^1D_2$,$^3P_0$,$^3P_1$, and $^3P_2$ where the triplet states have been split by the spin-orbit interaction. The other terms ^3D, ^3S, and ^1P, which are expected to be present by the vector coupling method, are really absent[3,10] in our case of two equivalent p electrons because of the Pauli exclusion principle. However, if the electrons are not equivalent as in the configuration npn′p ($n \neq n'$), such terms would be present.

The LS terms arising from various sets of equivalent electrons have been derived by Gibbs et al.[195] In Table 18 we have listed the LS terms for a few configurations of equivalent electrons.

If more than two electrons are present in outer shells, the terms for these electrons can be determined by first obtaining the terms for any two electrons and then combining the resulting terms with the ℓ and s values of the third electron to derive the new terms, and so on until all the electrons are taken into account to form the final terms. For example, the LS terms for the configuration $3p^24s$ would be as follows:

$$^1S + s \rightarrow {}^2S_{\frac{1}{2}}$$

$$^1D + s \rightarrow {}^2D_{\frac{3}{2}},\ {}^2D_{\frac{5}{2}}$$

$$^3P + s \rightarrow {}^2P_{\frac{1}{2}},\ {}^2P_{\frac{3}{2}},\ {}^4P_{\frac{1}{2}},\ {}^4P_{\frac{3}{2}},\ {}^4P_{\frac{5}{2}}$$

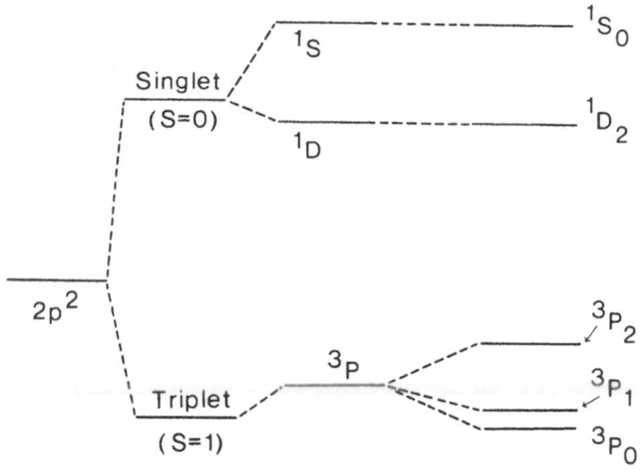

FIGURE 29. Fine-structure splitting of $2p^2$ electrons in the ground state of the carbon atom.

Table 18
LIST OF LS TERMS FOR A FEW SETS OF EQUIVALENT ELECTRONS

Configuration	Terms
s^2	^1S
p^1, p^5	^2P
p^2, p^4	^1S, ^1D, ^3P
p^3	^2P, ^2D, ^4S
d^1, d^9	^2D
d^2, d^8	^1S, ^1D, ^1G, ^3P, ^3F
d^3, d^7	^2D, ^2P, ^2D, ^2F, ^2G, ^2H, ^4P, ^4F
d^4, d^6	^1S, ^1D, ^1G, ^3P, ^3F, ^1S, ^1D, ^1F, ^1G, ^1I, ^3P, ^3D, ^3F, ^3G, ^3H, ^5D
d^5	^2D, ^2P, ^2D, ^2F, ^2G, ^2H, ^4P, ^4F, ^2S, ^2D, ^2F, ^2G, ^2I, ^4D, ^4G, ^6S

The excited configurations can be treated analogously. For example in the carbon atom, the lowest excited configuration is 2p3s, which generates the terms $^1P_1$, $^3P_0$, $^3P_1$, $^3P_2$. These states lie higher in energy by about 7 eV from the ground state term ^3P of the configuration $2p^2$. In Figure 30 we show some of the low-lying terms of the neutral carbon atom and also the allowed transitions (by dipole selection rules) which give rise to the experimental optical lines in the emission and absorption spectra. Similar term diagrams can be drawn from the tabulated values given by Moore[196] and Weiss et al.[197] for various atoms and ions.

It is important to remark that when a subshell is completely filled, the only allowed state corresponds to $s' = 0$, $\ell' = 0$, and $j' = 0$, i.e., the only state that occurs is $^1S_0$. Also, if a subshell is more than half filled, the sign of the spin-orbit interaction is reversed since now the holes (which behave like positively charged particles) play a part instead of the negatively charged electrons. Consequently, the energy level with maximum j' lies lowest; though maximum s' and maximum ℓ' still gives the lowest energy level since the sign of the residual Coulomb interaction is unaltered.

F. Zeeman Effect

The energy levels of atoms are split into a number of components in the presence

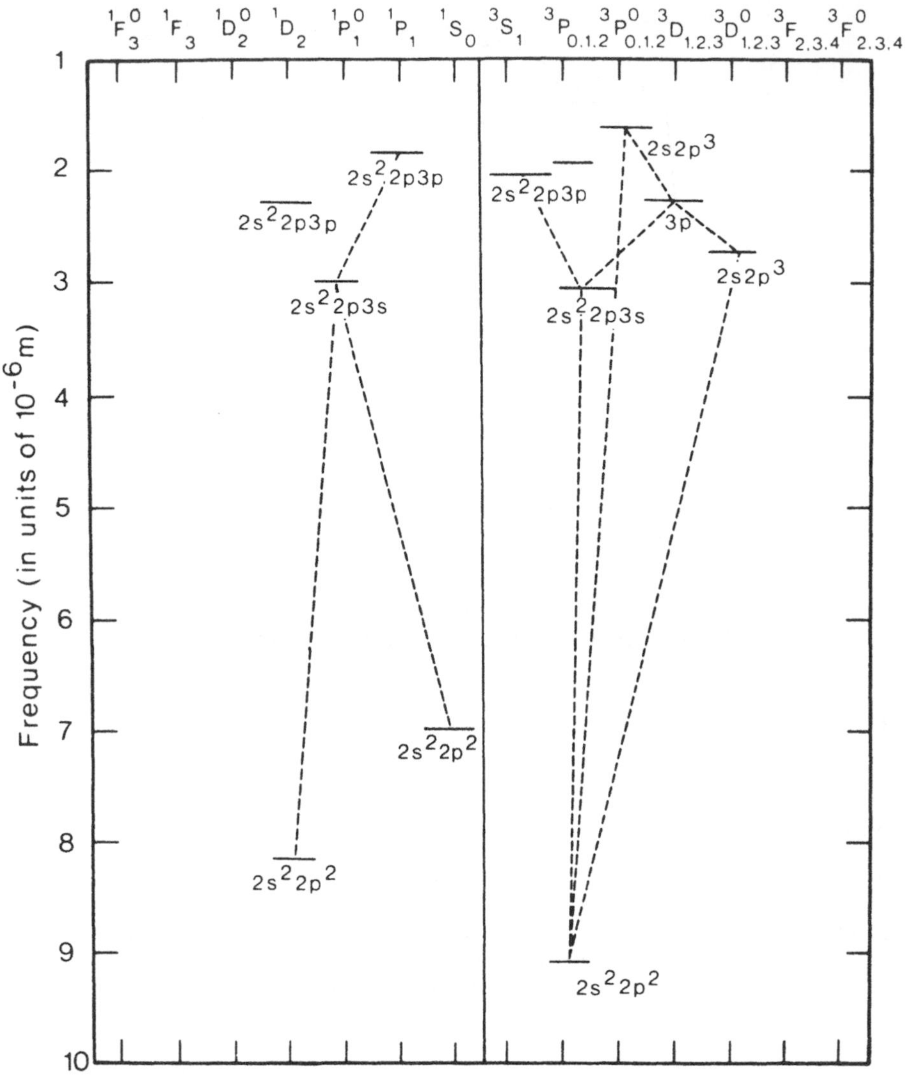

FIGURE 30. Some of the LS terms for neutral carbon and various allowed transitions.

of an applied magnetic field (also see Section II). This is the Zeeman effect discovered by Zeeman[29] in 1896. The Zeeman splitting is proportional to the magnetic field when the field is not very high in magnitude (< 0.1 T) and is smaller than the fine-structure splitting. Modern quantum theory has supplied a complete explanation for the Zeeman effect. The Zeeman effect results from the interaction of the net magnetic moment of the atom with the external magnetic field. The interaction energy is given by

$$W = -\vec{\mu} \cdot \vec{B}$$

where $\vec{B}$ is the applied magnetic field and $\vec{\mu}$ is the net magnetic moment of the atom which is the vector sum of the orbital and spin magnetic moments of the individual electrons:

$$\vec{\mu} = [\Sigma_i \frac{1}{2}\frac{e}{m}\vec{L}_i + \frac{e}{m}\vec{S}_i]$$

$$= -\frac{1}{2}\frac{e}{m}(\vec{J}' + \vec{S}')$$

$$= -\mu_B \frac{1}{\hbar}(\vec{J}' + \vec{S}') \qquad (316)$$

We note that $\vec{\mu}$ is not, in general, in the direction of $\vec{J}'$. In the absence of $\vec{B}$ the total angular momentum $\vec{J}'$ is a constant and $\vec{L}'$, $\vec{S}'$, and $\vec{\mu}$ precess about $\vec{J}'$. However, when $\vec{B}$ is non-zero, the total angular momentum $\vec{J}$ will no longer remain constant and will precess about $\vec{B}'$. For weak fields this precession will be much slower than the precession of $\vec{L}'$ and $\vec{S}'$ about $\vec{J}'$. In that case the time-averaged component of $\vec{\mu}$ along the field is about the same as the component of $\vec{\mu}$ along $\vec{J}'$ multiplied by the component of $\vec{J}'$ along the field. Accordingly,

$$W = -\frac{\mu_B}{\hbar}\frac{[(\vec{J}'+\vec{S}')\cdot\vec{J}'][\vec{J}'\cdot\vec{B}]}{J'^2} \qquad (317)$$

which gives the first-order energy shift,

$$\langle W \rangle = \Delta E = \mu_B B g m_{j'} \qquad (318)$$

where

$$m_{j'} = -j', -j'+1, \ldots, (j'-1), j' \qquad (319)$$

and

$$g = 1 + \frac{j'(j'+1) + s'(s'+1) - \ell'(\ell'+1)}{2j'(j'+1)} \qquad (320)$$

which is known as the Landé g-factor. The g-factor reduces to g_ℓ (i.e., equal to 1) if $s' = 0$ and it reduces to g_s (equal to 2) if $\ell' = 0$. Thus the Landé g-factor in Equation 320 is the general g-factor when the net angular momentum is partly orbital and partly spin. It is obvious from Equations 318 and 319 that the energy level is split into its $(2j' + 1)$ components corresponding to $(2j' + 1)$ values of m_j'. In Figure 31 we have depicted the Zeeman splittings of sodium levels $^2S_{1/2}$, and $^2P_{3/2}$ for which the calculations from Equation 320 yield the g-values 2, ⅔, and 4/3, respectively.

The selection rule for the transition between the Zeeman split levels is given by

$$\Delta m_{j'} = 0, \pm 1 \quad (m_{j'} = 0 \leftrightarrow m_{j'} = 0 \quad \text{if } \Delta_{j'} = 0) \qquad (321)$$

The Zeeman splittings are very useful for determining the ℓ', s′, j′ assignments of the various levels of an atom and thus confirming whether the atom obeys LS coupling or not.

If the applied magnetic field is stronger than the internal magnetic field which couples $\vec{S}'$ and $\vec{L}'$ to form $\vec{J}'$, the $\vec{L}'\vec{S}'$ coupling is destroyed. In that situation $\vec{S}'$ and $\vec{L}'$ precess independently about $\vec{B}$ and the orbital magnetic moment $\vec{\mu}$

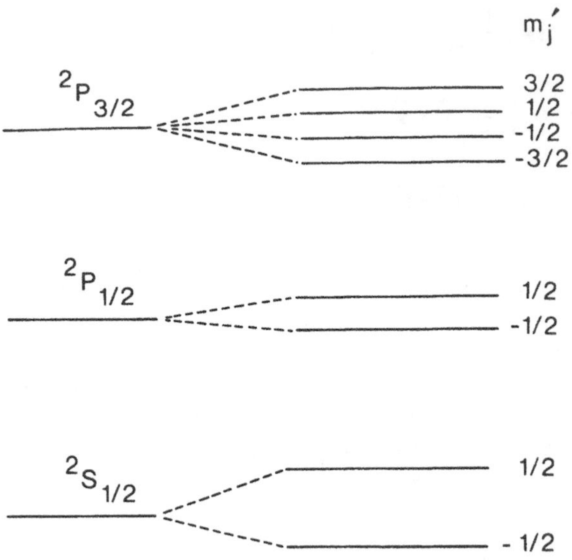

FIGURE 31. Schematic representation of the Zeeman splitting of the $^2S_{1/2}$, ^2P, and $^2P_{3/2}$ levels of Na.

$$\vec{\mu} = -\frac{\mu_B}{\hbar}(\vec{L}' + 2\vec{S}') \qquad (322)$$

couples with $\vec{B}$ to give the first-order interaction energy

$$\Delta E = \mu_B B(m_{\ell'} + 2m_{s'}) \qquad (323)$$

This is known as the Paschen-Bach effect. In this case the dipole selection rules are

$$\Delta m_{s'} = 0 \qquad (324)$$

$$\Delta m_{\ell'} = 0, \pm 1 \qquad (325)$$

for the transition between the split energy levels. The explanation for these selection rules is similar to that discussed in Section II.

V. THE AUGER EFFECT

A. Introduction

We have seen that if there is a vacancy in the K level of an atom, an electron from the L level (of the atom) transfers to the K level emitting a K_α X-ray photon. In fact, characteristic X-ray photons of different frequencies are emitted when electrons fall from the higher levels to the inner level vacancies of the atom, thereby constituting the fluorescent spectra of the atom. The electron emitted from the K shell by the direct action of an electromagnetic radiation is known as the K photoelectron. Auger[198] discovered, in a cloud chamber investigation of photoelectrons, that an atom which has a vacancy in an inner shell can return to the de-excited state by simultaneously dropping an electron from the higher shell into the vacant shell and ejecting another electron from the atom, usually from the same higher shell. This phenomenon is known as the Auger effect, and is different from the phenomenon of radiative de-excitation in which a photon emerges in the de-excitation process. In the Auger process, the energy lost by the electron which falls from the higher level to the lower level vacancy, is utilized in ejecting the other electron from the atom. It must be stressed that the second electron (the Auger electron) is not ejected by the process of the photoelectric absorption of the photon emitted when the first electron drops to the vacant state, but by the direct process of the readjustment of the atom. For instance, the initial state of the atom may correspond to a hole in the K level and the final state to holes in the L_I and L_{III} levels; the corresponding Auger transition may be denoted as a KL_IL_{III} transition. The energy ΔE acquired by the ejected electron is equal to the energy difference of the initial and the final states of the atom; that is, for the above example,

$$\Delta E = E_K - E_{L_I} - E_{L_{III}}$$

where E_K, E_{L_I} and $E_{L_{III}}$ are the energies of the K, L_I, and L_{III} atomic states. The probability of occurrence of the Auger transition is very high and is comparable to that of the radiative transition. In the process of readjustment of the atom there could also be the ejection of two or more electrons by multiple Auger transitions which leads to what is known in optical spectroscopy as the autoionization of the atom.

There are several Auger transitions depending on the initial vacancy and the final vacancies in the shells of an atom. Thus, besides the KL_IL_{II} Auger transition, transitions such as KL_iM_j are possible, in which the initial vacancy is in the K shell and the final vacancies are in L_i and M_j shells where i = I, II, III, and j = I, II, III, IV, V.

Thus we have seen that the creation of a vacancy in the inner shell initiates a complicated rearrangement of the electrons in the atom. Such processes take place in a time of the order of 10^{-17} to 10^{-14} sec and give rise to X-ray emission and Auger electrons. The mechanisms for these processes are well understood. The radiative transitions occur via electric multipole modes (mainly, the electric dipole), whereas the nonradiative or Auger transitions occur by virtue of the Coulomb interactions between the electrons.[199] For detailed calculations of the nonradiative transition rates one may refer to the articles by Asaad,[200,201] Listengarten,[202,203] Dionisio,[204] and Callan.[205]

Though a certain care is needed in defining the probabilities of emission of radiative and nonradiative processes, the recent definitions of the fluorescence yields, ω_i, and Auger yields, a_i, of the ith subshell of an atom may be given as follows. The probability that a vacancy in the ith subshell is filled by a radiative transition is ω_i (the fluorescence yield), whereas the probability that an electron is emitted when the vacancy in the ith subshell is filled by an electron from the higher shell is a_i, the Auger yield.

The K shell has only a single subshell and therefore the definitions of the fluorescence yields ω_i and the Auger yields a_i are straightforward and simple in nature. However, complications arise in defining the yields for higher atomic shells. For example, there are three L subshells, namely, L_I, L_{II}, and L_{III} in an atom and the number of vacancies in these subshells and the way in which these subshells are ionized bear on the average fluorescence yields since, in general, the individual fluorescence yield of the subshells (ω_{L_I}, $\omega_{L_{II}}$, and $\omega_{L_{III}}$) are not equal. Also, another complicating factor is that (for a certain range of atoms) the vacancy moves from one L subshell to the other by a nonradiative process (i.e., the electron motions $L_I \rightarrow L_{III}$, $L_I \rightarrow L_{II}$, or $L_{II} \rightarrow L_{III}$) before it is filled by an electron from the upper shell (M, N, etc.). Such transitions are known as Coster-Kronig[206] transitions. Experimentally, one measures the average fluorescent yield $\overline{\omega}_L$, which is the average of the subshell fluorescent yields ω_{L_I}, $\omega_{L_{II}}$, and $\omega_{L_{III}}$; these subshell yields must be adequately accounted for in theoretical interpretations.

More complicated situations[204] arise when one is concerned with even higher shells since they contain more subshells (for instance, the M shell has five subshells). To date very scant experimental data are available for M or higher shells and therefore further experimental efforts in this direction are warranted.

The radiationless processes mentioned above play an important role not only in atomic physics but also in molecular, nuclear, elementary particle, solid state, and surface physics. Several excellent reviews[207-212] are now available on this subject.

Numerous experimental studies[200,213-219] have been made for obtaining the precise Auger spectrum, in many cases using high resolution electron spectrometers. It is now possible to accurately measure even very weak electron lines through improved techniques. Also, theoretical calculations have been performed by many researchers.[220-222] While the broad features of the experimental K Auger spectra have been confirmed by theoretical calculations incorporating relativistic effects and configuration interactions, the detailed agreement between the theory and the experiments has not yet been realized because of the inaccuracies in the inner shell wave functions. The L and M Auger spectra, because of their complicated structure, still await careful interpretations.

The radiative Auger effect, as predicted by Bloch,[223] had also been observed by Aberg and Utriainen[224] and has been discussed by Burhop and Asaad.[210] This effect is equivalent to an internal Compton process in which an X-ray photon is emitted when an electron jumps into the K shell vacancy and ejects another electron from the higher shell.

B. Theory of the Auger Effect

1. Introduction

In the following we shall treat separately the processes which govern the radiationless transition and the processes responsible for the radiative transition and primary ionization in atoms. The theory of transition rates for the radiationless transition and the radiative transition rates will be reviewed briefly. Besides the transition rates, other quantities of interest are the fluorescence yields and Auger yields associated with various transitions.

The Auger spectrum, in general, is very complex and contains many lines which pose difficulties for their analysis. In this circumstance the fluorescence yields, as determined from experiments, serve a useful purpose to furnish information about the details of the Auger effect. The theoretical estimates of the fluorescence yields, on the other hand, demonstrate the progress of the relevant theory and the degree of accuracy of the calculated wave functions.

2. Auger Transition Rates

We wish to present here a brief account of the nonrelativistic theory of the Auger effect. For a review of the relativistic treatment one may refer to the article by Burhop and Asaad.[210] The nonrelativistic theory was developed first by Wentzel[225] and has been reviewed by Burhop,[207] by Bergstrom and Nodling,[199] and by Burhop and Asaad.[210]

Our main concern is with the nonradiative change of states of two electrons that are initially bound to an atom; in their final states one electron is bound to the atom whereas the other electron is free in the continuum. In the whole process the total energy is unchanged. The nonrelativistic Hamiltonian for the two electrons may be written as

$$\mathcal{H} = -\frac{\hbar}{2m}(\nabla_1^2 + \nabla_2^2) + V(1) + V(2) + \frac{e^2}{4\pi\epsilon_0}\frac{1}{r_{12}} \quad (326)$$

Here V(1) and V(2) are the potentials seen by the two electrons due to the nucleus and the average effect of the remaining electrons in the atom. Let the initial and final states of the electrons be represented by the functions Ψ_i (1,2) and Ψ_f(1,2), respectively. Then, assuming[226] the Coulomb interaction between the two electrons (the last term in Equation 326) as a perturbation, the radiationless (Auger) transition rate b_A from the initial state to the final state is given by time-dependent first-order perturbation theory as[226]

$$b_A = \frac{2\pi}{\hbar} \left| \iint \Psi_f^*(1,2) \frac{e^2}{4\pi\epsilon_o r_{12}} \Psi_i(1,2)\, d^3r_1\, d^3r_2 \right|^2 \quad (327)$$

If the electrons 1 and 2 are initially in the bound states described by Ψ_i and Ψ_i', and finally in the states Ψ_f and Ψ'_f where Ψ_f represents the bound state and Ψ_f' the continuum state, the total antisymmetric and normalized wave functions may be expressed as

$$\Psi_i(1,2) = \frac{1}{\sqrt{2}} [\psi_i(1)\, \psi_i'(2) - \psi_i(2)\, \psi_i'(1)] \quad (328)$$

$$\Psi_f(1,2) = \frac{1}{\sqrt{2}} [\psi_f(1)\, \psi_f'(2) - \psi_f(2)\, \psi_f'(1)] \quad (329)$$

Making use of Equations 328 and 329 in Equation 327, one obtains the Auger transition rate,

$$b_A = \frac{2\pi}{\hbar} |R_A^{(d)} - R_A^{(e)}|^2 \quad (330)$$

where $R_A^{(d)}$ and $R_A^{(e)}$ are the direct and exchange transition amplitudes defined by

$$R_A^{(d)} = \iint \psi_f^*(1)\, \psi_f'^*(2) \frac{e^2}{4\pi\epsilon_o r_{12}} \psi_i(1)\, \psi_i'(2)\, d^3r_1\, d^3r_2 \quad (331)$$

and

$$R_A^{(e)} = \iint \psi_f^*(1)\, \psi_f'^*(2) \frac{e^2}{4\pi\epsilon_o r_{12}} \psi_i(2)\, \psi_i'(1)\, d^3r_1\, d^3r_2 \quad (332)$$

In the general case one is required to consider the N-electron Hamiltonian in lieu of Equation 326 for a N-electron atom ($N \leqslant Z$). Following the treatment given by Burhop and Asaad,[210] the N-electron Hamiltonian is

$$\mathcal{H} = -\frac{\hbar^2}{2m}\sum_{i=1}^{N} \nabla_i^2 + \sum_{i=1}^{N} V(i) + \sum_{i<j} \left\{ \frac{e^2}{4\pi\epsilon_o r_{ij}} - \left(\overline{\frac{e^2}{4\pi\epsilon_o r_{ij}}} \right) \right\} \quad (333)$$

where $\overline{(e^2/4\pi\varepsilon_o r_{ij})}$ stands for the average value of the potential of the electrons. Constructing now the Slater determinants for the total initial and the final states as

$$\Psi_i(1,2,\ldots N) = \frac{1}{\sqrt{N!}} \mathrm{Det}(\psi_i^{\alpha}(1)\ \psi_i^{\beta}(2)\ldots\psi_i^{\nu}(N)) \tag{334}$$

$$\Psi_f(1,2,\ldots N) = \frac{1}{\sqrt{N!}} \mathrm{Det}(\psi_f^{\alpha}(1)\ \psi_f^{\beta}(2)\ldots\psi_f^{\nu}(N)) \tag{335}$$

and using the simplifications familiar in the case of atomic structure calculations,[11,227] one obtains the same expression as given in Equation 330 for the Auger transition rate.

The process considered above may be described in an alternate way that deals with the hole states instead of the electron states. Let the initial states be identified in terms of the usual atomic quantum numbers as

$$\psi_i \equiv \psi_{n\ell m_\ell m_s} \tag{336}$$

$$\psi_i' \equiv \psi_{n'\ell' m_\ell' m_s'} \tag{337}$$

and the final bound state as

$$\psi_f \equiv \psi_{n''\ell'' m_\ell'' m_s''} \tag{338}$$

Then the process in terms of the hole states can be expressed by the transition

$$(n''\ell'' m_\ell'' m_s'') \rightarrow (n\ell m_\ell m_s)\ (n'\ell' m_\ell' m_s') \tag{339}$$

where it is clear that the system undergoes the transition from the initial state characterized by the hole state $n''\ell''m_\ell''\ m_s''$ to the final state characterized by the hole states $(n\ell m_\ell m_s)$ and $(n'\ell'm'_\ell m'_s)$. Next, one sums the Auger transition rates for all possible $m_\ell m'_\ell m_s\ m_s'\ m_\ell''\ m_s''$ values to obtain the net probability of the Auger transition, i.e., $W[n''\ell'' \rightarrow (n\ell)\ (n'\ell')]$, which is proportional to the intensity of the Auger electrons. Thus, in the notation of Equation 339, the various Auger transitions can be expressed by $K \rightarrow L_I L_{II}$, $K \rightarrow L_I l_{III}$, and so on. One notes from Equations 320, 331, 332, 336, 337, and 338 that the Auger transition consists of two different processes, namely, direct and exchange. These processes have been depicted schematically in Figure 32.

We have followed a nonrelativistic derivation of the Auger transition rate in the above. However, the importance of the relativistic effects is well known for the heavy atoms. Several relativistic calculations of the Auger transition probabilities have been performed by many researchers and have been reviewed by Burhop and Asaad.[210]

3. Radiative Transition Rates

The nonrelativistic radiative transition rate can be derived from the first-order perturbation theory by considering the interaction of the electromagnetic wave with the electron in the atom which undergoes a transition from the initial state Ψ_i to the final state Ψ_f. From Section II (Equations 193 to 198), the radiative transition rate can be written as

$$b_r = \frac{8\pi^3 \nu_{if}^3}{3\pi \epsilon_o \hbar c^3} |\vec{M}|^2 \tag{340}$$

where $\vec{M}$ is the electric dipole moment matrix element,

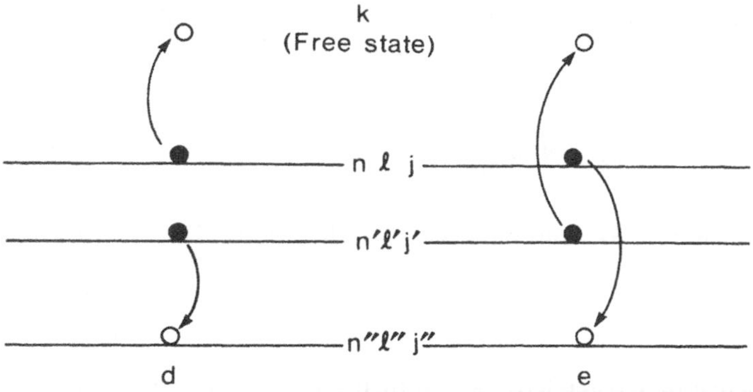

FIGURE 32. Direct (d) and exchange (e) Auger processes have been shown schematically. Empty circles show the vacancies whereas the filled circles are the electrons. The transitions have been depicted by the arrows.

$$\vec{M} = e \int \Psi_f^* \, \vec{r} \, \Psi_i \, d^3r \tag{341}$$

and ν_{if} is the frequency of the radiation emitted corresponding to the transition from the initial state to the final state.

For the relativistic calculations of the radiative rates, particularly for K and L holes, one may consult the articles by Scofield[228] (who considered multipole radiations and the effect of retardation) and others.[229-231] It has been found that the radiative transition rates are lowered by relativistic effects by about 25%, as in case of gold.[230,231]

4. The Fluorescence Yield

The fluorescence yield ω_i for the ith subshell of an atom is defined by

$$\omega_i = \frac{P_r^i}{P_r^i + P_A^i} \tag{342}$$

where P_r^i is the net transition rate for all possible radiative transitions of the type expressed in Equation 340 to fill the vacancy in the ith shell, and P_A^i is the total nonradiative transition rate. (See Equation 330 for all possible Auger transitions for shell i.)

The expressions Equations 330 and 340 given above are used for estimating the total transition probabilities P_A^i and P_R^i and hence, the fluorescence yields from Equation 342.

5. Various Symbols and Relations

The fluorescence yield for the K shell is denoted by ω_k. For the L shell one defines the average fluorescence yield $\bar{\omega}_L$ by

$$\bar{\omega}_L = N_1 \omega_{L_I} + N_2 \omega_{L_{II}} + N_3 \omega_{L_{III}} \tag{343}$$

where N_1, N_2, and N_3 are the relative number of vacancies in the subshells, L_I, L_{II}, and L_{III}, respectively, and ω_i are the individual fluorescence yields for the subshells i. Obviously,

$$N_1 + N_2 + N_3 = 1 \tag{344}$$

For the different relative number of vacancies one can write expressions analogous to Equation 343. Thus for the primary vacancies N_1', N_2', N_3', we have the corresponding average fluorescence yield,

$$\overline{\omega}_L' = N_1'\omega_{L_I} + N_2'\omega_{L_{II}} + N_3'\omega_{L_{III}} \tag{345}$$

and, similarly, for N_1'', N_2'', and N_3'' vacancies, the average fluorescence yield is

$$\overline{\omega}_L'' = N_1''\omega_{L_I} + N_2''\omega_{L_{II}} + N_3\omega_{L_{III}}'' \tag{346}$$

In the presence of Coster-Kronig[206] transitions, the situation is more complicated and requires modification of the above equations.

Following Wapstra et al.,[232] the modified equation for $\overline{\omega}_L$ becomes

$$\overline{\omega}_L = V_1\omega_{L_I} + V_2\omega_{L_{II}} + V_3\omega_{L_{III}} \tag{347}$$

where V_1, V_2, and V_3 are the modified vacancy distributions in the presence of the Coster-Kronig processes. Since some of the vacancies created in L_{II} and L_{III} subshells may be counted twice,

$$V_1 + V_2 + V_3 \geqslant 1 \tag{348}$$

Next we define the Coster-Kronig transition probabilities f_{12}, f_{13}, and f_{23} as the probabilities for the Coster-Kronig transitions $L_I \rightarrow L_{II}$, $L_I \rightarrow L_{III}$, and $L_{II} \rightarrow L_{III}$, respectively. Then the vacancy distributions V_1, V_2, and V_3 are related to the primary distributions by

$$V_1 = N_1$$

$$V_2 = N_2 + f_{12}N_1$$

$$V_3 = N_3 + f_{23}N_2 + (f_{13} + f_{12}f_{23})N_1 \tag{349}$$

It is now possible to express $\overline{\omega}_L$ in terms of the primary vacancy distributions N_1, N_2, and N_3 as

$$\overline{\omega}_L = N_1\nu_{L_I} + N_2\nu_{L_{II}} + N_3\nu_{L_{III}} \tag{350}$$

where ν_{L_I} is the new "fluorescence yield," defined such that it represents the fraction of all L X-rays emitted per L_I subshell vacancy. $\nu_{L_{II}}$ and $\nu_{L_{III}}$ carry similar meanings. With the help of Equations 347, 349, and 350 one immediately obtains the relations connecting ν'_is with ω'_is. Thus,

$$\nu_{L_I} = \omega_{L_I} + (f_{13} + f_{12} + f_{23})\,\omega_{L_{III}}$$

$$\nu_{L_{II}} = \omega_{L_{II}} + f_{23}\,\omega_{L_{III}}$$

$$\nu_{L_{III}} = \omega_{L_{III}} \tag{351}$$

As mentioned previously, the fluorescence yield ω_i is the probability that a vacancy in the ith subshell is filled with an electron by a radiative transition. Also, the Auger yield a_i is the probability that the vacancy in the ith shell is filled with an electron by a nonradiative transition from a higher shell, with the assumption that Coster-Kronig transitions are absent. The Coster-Kronig yield is defined as the probability that a vacancy is filled with an electron by a nonradiative process from the higher subshell in the same primary shell.

The above definitions lead to the relations

$$\omega_{L_{III}} + a_{L_{III}} = 1$$

$$\omega_{L_{II}} + a_{L_{II}} + f_{23} = 1$$

$$\omega_{L_I} + a_{L_I} + f_{12} + f_{13} = 1 \tag{352}$$

As for the average L-shell Auger yield, $\bar{a}_L$, one writes[233] (in analogy with the above treatment see Equation 347)

$$\bar{a}_L = a_{L_I}V_1 + a_{L_{II}}V_2 + a_{L_{III}}V_3 \tag{353}$$

The above expressions (Equations 347, 349, and 353) readily lead to

$$\bar{a}_L + \bar{\omega}_L = 1$$

That is, the sum of the average fluorescence yield and the Auger yield is unity, as expected.

C. Results for Fluorescence Yields and Coster-Kronig Yields

The Auger transition rates have been calculated by various investigators.[200,234-247] Burhop[247] obtained K → LL transition rates for Ag using the screened hydrogenic wavefunctions. Pincherle[246] has made similar calculations incorporating interactions from higher shells but using only the unscreened hydrogenic wave functions. The wave functions obtained by the Hartree self-consistent field method have been used by Rubenstein[239] and Asaad[200] to estimate the various Auger transitions entailing the cases of Ar, Kr, Ag, and Hg. Callen,[243] Callen et al.,[242] and Kostroun et al.[241] have performed calculations by improved methods using the screened hydrogenic functions. The Hartree-Fock-Slater wave functions (see also Section III) have been employed by McGuire[237,238] and Walters and Bhalla[235,236] to obtain the Auger transition rates for many systems. Among earlier calculations Ramberg and Richtmyer[244] have used the Thomas-Fermi model for investigating the K → LL transition of gold.

The radiative transition rates have been estimated by McGuire,[237,238] Asaad,[200] Rubenstein,[239] Pincherle,[246] Gaffrion and Nadeau,[245] and others. McGuire[248] has evaluated radiative matrix elements for several atoms for an initial vacancy in the L state. In these references either unscreened hydrogenic wave functions, Hartree self-consistent wave functions, or Herman-Skillman wave functions have been employed.

Relativistic L-shell Auger and Coster-Kronig ratio and fluorescence yields have recently been calculated by Chen et al.[249] for selected atoms with $70 \leqslant Z \leqslant 96$ using frozen orbitals in the Dirac-Hartree-Slater approximation. The earliest relativistic treatment of Auger transitions is by Massey and Burhop[231] who deduced K-LL rates associated with gold from screened hydrogenic wave functions. Other calculations include those of Asaad,[200] Listengarten,[203] Bhalla and Ramsdale,[250] Chatterji and Talukdar,[251] and others.

The values of the fluorescence yields derived from experiments and from calculations of the Auger and radiative transition rates have been discussed by Burhop and Asaad,[210] Bambynek et al.,[212] Wapstra et al.,[232] Fink et al.,[209] and McGuire.[211]

We have listed values of K shell fluorescent yields ω_k in Table 19 and values of L subshell fluorescent yields ω_{L_I}, $\omega_{L_{II}}$ and $\omega_{L_{III}}$ in Table 20. The values of the Coster-Kronig yields f_{12}, f_{13}, and f_{23} have been tabulated in Table 21. The average values of the L fluorescent yields $\overline{\omega}_L$ have been given in Table 22. A plot of ω_K as a function of the atomic number has been shown in Figure 33.

Table 19
LIST OF EXPERIMENTAL AND THEORETICAL K-FLUORESCENT YIELDS ω_K FOR VARIOUS ATOMS

Atomic number	Element	ω_K Experiment	ω_K Theory
4	Be	0.000304	
5	B	0.00071	
6	C	0.0009	
7	N	0.0015	
8	O	0.0022	
10	Ne	0.043	
12	Mg	0.028	
13	Al	0.038	
14	Si	0.038	
16	S	0.083	0.098
17	Cl	0.093	0.117
18	Ar	0.087	0.13
19	K		0.155
20	Ca	0.15	0.165
22	Ti	0.22	0.213
23	V	0.24	0.242
24	Cr	0.277	0.272
25	Mn	0.273	0.291
26	Fe	0.308	0.319
27	Co	0.31	0.346
28	Ni	0.38	0.378
29	Cu	0.454	0.407
30	Zn	0.430	0.438
31	Ga	0.427	0.469
32	Ge		0.510
33	As	0.53	0.548
34	Se	0.578	0.585
35	Br	0.565	0.628
36	Kr	0.660	0.66
37	Rb		0.680
38	Sr	0.64	0.702
39	Y		0.719
40	Zr	0.70	0.737
41	Nb	0.73	0.754
42	Mo	0.73	0.770
43	Tc	0.697	0.785
44	Ru		0.799
45	Rh	0.786	0.812

Table 19 (continued)
LIST OF EXPERIMENTAL AND THEORETICAL K-FLUORESCENT YIELDS ω_K FOR VARIOUS ATOMS

		ω_K	
Atomic number	Element	Experiment	Theory
46	Pd	0.790	0.822
47	Ag	0.821	0.84
48	Cd	0.827	0.843
49	In	0.820	
50	Sn	0.846	
51	Sb	0.862	
52	Te	0.872	
53	I	0.91	
54	Xe	0.88	0.872
55	Cs	0.898	
56	Ba	0.85	
57	La	0.94	
58	Ce	0.90	
59	Pr	0.90	
63	Eu	0.908	
64	Gd	0.925	
65	Tb		0.924
66	Dy	0.943	
67	Ho		0.930
68	Er	0.955	
69	Tm		0.936
70	Yb	0.936	
71	Lu		0.944
73	Ta		0.946
74	W		0.945
75	Re		0.951
77	Ir		0.956
78	Pt	0.942	
79	Au		0.953
80	Hg	0.952	0.955
81	Tl		0.962
82	Pb	0.96	0.963
83	Bi	0.96	0.963
84	Po	0.944	0.963
90	Th		0.963
91	Pa		0.963
92	U	0.967	0.963
93	Np	0.938	0.963
94	Pu		0.964
96	Cm		0.965
98	Cf		0.966
100	Fm		0.967

Table 20
TABULATION OF L SUBSHELL FLUORESCENT YIELDS, ω_{L_I}, $\omega_{L_{II}}$, AND $\omega_{L_{III}}$ FOR VARIOUS ELEMENTS

Atomic number (Z)	Element	ω_{L_I}	$\omega_{L_{II}}$	$\omega_{L_{III}}$
37	Rb		0.0129	0.0152
38	Sr		0.0173	0.0165
39	Y		0.0196	0.0196
40	Zr		0.0188	0.0221
41	Nb		0.0245	0.0242
42	Mo		0.0254	0.0271
44	Ru		0.0360	0.0374
45	Rh		0.0391	0.0383
46	Pd		0.0448	0.0420
47	Ag		0.0427	0.0453
48	Cd		0.0468	0.0453
49	In		0.0505	0.0536
50	Sn		0.0441	0.0579
56	Ba			0.05
65	Tb		0.165	0.188
67	Ho		0.170	0.169
68	Er		0.185	0.172
70	Yb		0.188	0.183
71	Lu		0.257	0.251
72	Hf		0.299	0.228
73	Ta		0.257	0.228
			0.270	0.254
74	W		0.295	0.272
75	Re		0.310	0.284
76	Os		0.328	0.290
77	Ir		0.317	0.262
78	Pt		0.341	0.317
79	Au		0.354	0.317
80	Hg	0.082	0.408	0.320
			0.319	0.300
81	Tl		0.400	0.386
			0.319	0.306
82	Pb	0.11	0.363	0.315
		0.07		
		0.09		
83	Bi		0.38	0.340
88	Ra		0.472	
90	Th		0.473	0.517
			0.44	
91	Pa		0.55	0.46
92	U	0.15	0.545	0.443
			0.529	
93	Np	0.19	0.78	0.57
94	Pu		0.523	
			0.42	
96	Cm	0.18	0.552	0.477

Table 21
TABULATION OF L SUBSHELL COSTER-KRONIG FLUORESCENT YIELDS f_{12}, f_{13}, and f_{23} FOR VARIOUS ATOMS

Atomic number (Z)	Element	f_{12}	f_{13}	f_{23}
56	Ba			0.26
70	Yb			0.143
73	Ta		0.19	0.20
74	W		0.27	0.139
77	Ir		0.46	
78	Pt		0.50	
79	Au	0.25	0.51	0.22
			0.61	
80	Hg	0.069	0.705	0.127
				0.22
				0.08
81	Tl		0.76	
			0.57	
		0.17	0.56	0.25
81—83		0.24	0.52	0
82	Pb	0.16	0.60	0
83	Bi	0.16	0.62	0
		0.19	0.58	0.06
		0.19	0.58	0
88	Ra			0.110
90	Th			0.106
92	U	0.051	0.656	0.138
93	Np	0.10	0.55	0.02
96	Cm	0.047	0.629	0.211

Table 22
AVERAGE L SHELL FLUORESCENT YIELDS, $\overline{\omega}_L$

Atomic number	Element	$\overline{\omega}_L$
23	V	0.00235
25	Mn	0.00295
29	Cu	0.0056
31	Ga	0.0061
36	Kr	0.13
37	Rb	0.011
40	Zr	0.057
42	Mo	0.067
47	Ag	0.047, 0.029, 0.100
51	Sb	0.119
52	Te	0.073, 0.071, 0.122
54	Xe	0.103, 0.11, 0.21, 0.25
56	Ba	0.148
57	La	0.099, 0.092, 0.158
58	Ce	0.163
59	Pr	0.167
60	Nd	0.170
62	Sm	0.188
63	Eu	0.17

Table 22 (continued)
AVERAGE L SHELL FLUORESCENT YIELDS, ω_L

Atomic number	Element	ω_L
64	Gd	0.198
66	Dy	0.14
68	Er	0.228
72	Hf	0.260, 0.17
74	W	0.298
76	Os	0.348
78	Pt	0.348, 0.32
79	Au	0.365, 0.430
80	Hg	0.410, 0.34, 0.41, 0.40, 0.371, 0.24, 0.39
81	Tl	0.32, 0.50, 0.41, 0.48
82	Pb	0.39, 0.398, 0.36
83	Bi	0.38, 0.40, 0.402, 0.51
88	Ra	0.40, 0.52
91	Pa	0.52
92	U	0.603, 0.45
93	Np	0.66
94	Pu	0.486

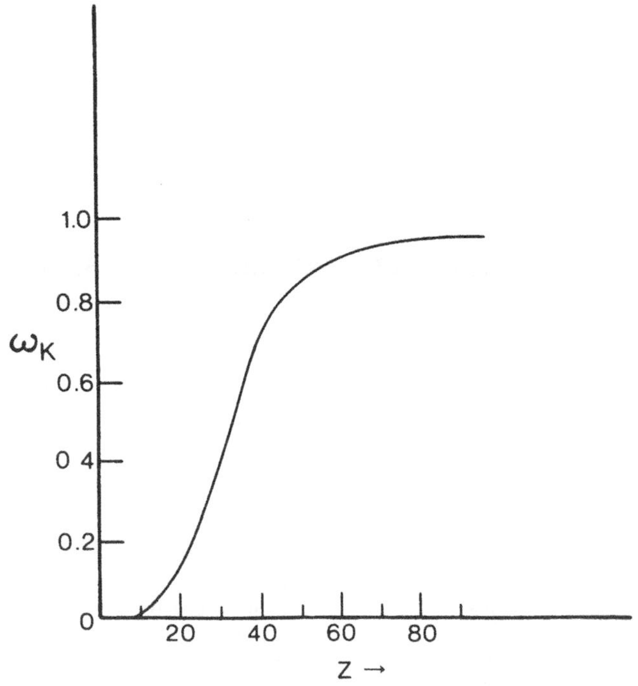

FIGURE 33. Plot of the experimental fluorescent yield ω_K as a function of the atomic number Z.

VI. CONCLUDING REMARKS

Starting with the basic principles of quantum mechanics we have described the theory of the one-electron and multielectron atoms and the various relevant phenomena. The development presented here is also helpful in explaining the fundamentals of quantum mechanics. We have kept the subject matter as simple as possible and have given many figures and tables to make it very transparent. However, some of the topics have been described very briefly and for these topics we have tried to give important pertinent references so that a reader may have easy access to the literature to gather advanced knowledge. It is hoped that this article will be useful not only to beginners in physics, chemistry, biology, and medicine, but also to others for consultation and reference purposes.

ACKNOWLEDGMENTS

Thanks are due to Dr. S. Sundaram and Mrs. M. H. de Viccaro for assisting the author in reading the manuscript and making appropriate corrections.

REFERENCES

1. **Mizushima, M.**, *Quantum Mechanics of Atomic Spectra and Atomic Structure*, W. A. Benjamin, New York, 1970.
2. **Anderson, A. A.**, *Modern Physics and Quantum Mechanics*, W. B. Saunders, Philadelphia, 1971.
3. **Leighton, R. B.**, *Principles of Modern Physics*, McGraw-Hill, New York, 1959.
4. **White, R. L.**, *Basic Quantum Mechanics*, McGraw-Hill, New York, 1966.
5. **Weidner, R. T. and Sells, R. L.**, *Elementary Modern Physics*, Allyn and Bacon, Boston, 1973.
6. **Morrison, M. A., Estle, T. L., and Lane N. F.**, *Quantum States of Atoms, Molecules and Solids*, Prentice-Hall, Englewood Cliffs, N.J., 1976.
7. **Hameka, H. F.**, *Introduction to Quantum Theory*, Harper & Row, New York, 1967.
8. **Saxon, D. S.**, *Elementary Quantum Mechanics*, Holden-Day, San Francisco, 1968.
9. **Stehle, P.**, *Quantum Mechanics*, Holden-Day, San Francisco, 1966.
10. **Slater, J. C.**, *Quantum Theory of Atomic Structure*, Vol. 1, McGraw-Hill, New York, 1960.
11. **Slater, J. C.**, *Quantum Theory of Atomic Structure*, Vol. 2, McGraw-Hill, New York, 1960.
12. **Condon, E. U. and Shortley, G. H.**, *Theory of Atomic Spectra*, Cambridge University Press, London, 1959.
13. **Griffith, J. S.**, *Theory of Transition Metal Ions*, Cambridge University Press, London, 1961.
14. **Eisberg, E. and Resnick, R.**, *Quantum Physics of Atoms, Molecules, Solids, Nuclei and Particles*, John Wiley & Sons, New York, 1974.
15. **Schiff, L. I.**, *Quantum Mechanics*, McGraw-Hill, New York, 1955.
16. **Merzbacher, E.**, *Quantum Mechanics*, John Wiley & Sons, New York, 1961.
17. **Pauling, L. and Wilson, E. B.**, *Introduction to Quantum Mechanics*, McGraw-Hill, New York, 1935.
18. **Rojansky, V.**, *Introduction to Quantum Mechanics*, Prentice-Hall, Englewood Cliffs, N.J., 1938.
19. **Park, D.**, *Introduction to Quantum Theory*, McGraw-Hill, New York, 1961.
20. **Eyring, E. M., Walter, J., and Kimball, G. E.**, *Quantum Chemistry*, John Wiley & Sons, New York, 1944.
21. **Bohm, D.**, *Quantum Theory*, Prentice-Hall, Englewood Cliffs, N.J., 1951.
22. **Dicke, R. H. and Wittke, J. P.**, *Introduction to Quantum Mechanics*, Addison-Wesley, Reading, Mass., 1960.
23. **Stern, O. and Gerlach, W.**, *Z. Phys.*, 8, 110, 1922; 9, 349, 1922.
24. **Phipps, P. E. and Taylor, J. B.**, *Phys. Rev.*, 29, 309, 1927.

25. **Lamb, W. E., Jr.** and **Rutherford, R. C.**, *Phys. Rev.*, 72, 241, 1947.
26. **Compton, A. H.** and **Allison, S. K.**, *X-Rays in Theory and Experiment*, D Van Nostrand, New York, 1935.
27. **Uhlenbeck, G. E.** and **Goudsmit, S. A.**, *Naturwissenschaften*, 13, 953, 1925; *Nature*, 117, 264, 1926; for a recent review see **Goudsmit, S. A.**, *Phys. Today*, p. 40, June, 1976.
28. **Dirac, P.A.M.**, *Proc. R. Soc. London, Ser. A*, 117, 610, 1928; see also **Dirac, P.A.M.**, *The Principles of Quantum Mechanics*, 4th ed., Oxford University Press, New York, 1958.
29. **Zeeman, P.**, *Philos. Mag.*, 5, 43, 226, 1897; *Phys. Z.*, 10, 217, 1909.
30. **Thomas, L. H.**, *Nature*, 117, 514, 1926.
31. **Russell, H. N.** and **Saunders, F. A.**, *Astrophys. J.*, 61, 38, 1925.
32. **Van Vlech, J. H.**, *The Theory of Electric and Magnetic Susceptibilities*, Oxford University Press, Oxford, 1932.
33. **Landau, L. D.** and **Liftshitz, E. M.**, *Quantum Mechanics*, Addison-Wesley, Reading, Mass., 1958.
34. **Sommerfeld, A.**, *Ann. Phys.*, 51, 1, 1916; **Wilson, W.**, *Philos. Mag.*, 29, 795, 1915.
35. **Feyman, R. P.**, *Quantum Electrodynamics*, W. A. Benjamin, New York, 1961; **Gupta, S. N.**, *Quantum Electrodynamics*, Gordon and Breach, New York, 1977.
36. **Drake, G. W. F., Goldman, S. P.**, and **van Wijngaarden, A.**, *Phys. Rev.*, A20, 1299, 1979.
37. **Lipworth, E.** and **Novick, R.**, *Phys. Rev.*, 108, 1434, 1957.
38. **Narasimham, M.** and **Strombtne, R.**, *Phys. Rev.*, A4, 14, 1971.
39. **Mohr, P. J.**, *Phys. Rev. Lett.*, 34, 1050, 1975; *Ann. Phys. (N.Y.)*, 88, 26, 1974; in *Beam-Foil Spectrocopy*, Sellin, I. A. and Pegg, D. J., Eds., Plenum Press, New York, 1976, 89.
40. **Erickson, G. W.**, *Phys. Rev. Lett.*, 27, 780, 1971; *J. Phys. Chem. Ref. Data*, 831, 1977.
41. **Pauli, W.** *Collected Scientific Papers*, Kronig, R. and Weisskopf, V. F., Eds., Interscience, New York, 1964.
42. **Slichter, C. P.**, *Principles of Magnetic Resonance*, Harper & Row, New York, 1963.
43. **Das, T. P.** and **Hahn, E. L.**, *Nuclear Quadrupole Resonance Spectroscopy*, Academic Press, New York, 1964.
44. **Armstrong, L., Jr.**, *Theory of Hyperfine Structure of Free Atoms*, 1st ed., John Wiley & Sons, New York, 1971.
45. **Szymanski, H. A., Ed.**, *Raman Spectrocopy*, Plenum Press, New York, 1967; **Suschinskii, M. M.**, *Raman Spectra of Molecules and Crystals*, Halsted Press, New York, 1972.
46. **Fermi, E.**, *Nuclear Physics*, University of Chicago Press, Chicago, 1950, 142.
47. **Judd, B. R.**, Selection rules within atomic shells, in *Advances in Atomic and Molecular Physics*, Vol. 7, 1st ed., Bates, D. R. and Easterman, I., Eds., Academic Press, New York, 1971, 252.
48. **Magnus, W.** and **Oberhetinger, F.**, *Formulas and Theorems for the Special Functions of Mathematical Physics*, Chelsea Publishing, New York, 1949.
49. **Heisenberg, W.**, *Z. Phys.*, 49, 619, 1928.
50. **Dirac, P. A. M.**, *Proc. R. Soc.*, A123, 714, 1929.
51. **Mattis, D. C.**, *The Theory of Magnetism*, Harper & Row, New York, 1965.
52. **Herzberg, G.**, *Atomic Spectra and Atomic Structure*, Dover Publication, New York, 1944.
53. **Morse, P. M.** and **Feshbach, H.**, *Methods of Theoretical Physics*, McGraw-Hill, New York, 1953, chap. 3.
54. **Moisewitch, B. L.**, *Variational Principles*, Interscience, New York, 1966.
55. **Gould, S. H.**, *Variational Methods for Eigenvalue Problems*, University of Toronto Press, Canada, 1966.
56. **Mikhli, S. G.**, *Variational Method in Mathematical Physics*, (Engl. trans.), Macmillan, New York, 1964.
57. **Yourgrau, W.** and **Mandelstam, S.**, Variational principles, in *Dynamics and Quantum Theory*, W. B. Saunders, Philadelphia, 1968.
58. For discussions on the bounds of energy eigenvalues one may consult **Lowdin, P. O.**, Perturbation theory and its applications, in *Quantum Mechanics*, Wilcox, C. H., Ed., John Wiley & Sons, New York, 1966.
59. **Schwartz, C.**, *Phys. Rev.*, 128, 1146, 1962.
60. **Pekeris, C. L.**, *Phys. Rev.*, 126, 1470, 1962; 127, 509, 1962.
61. **Schiff, B., Lifson, H., Pekeris, C. L.**, and **Rabinowicz, P.**, *Phys. Rev.*, 140, A1104, 1965.
62. **Pekeris, C. L.**, *Phys. Rev.*, 126, 143, 1962.
63. **Nesbet, R. K.** and **Watson, R. E.**, *Phys. Rev.*, 110, 1073, 1958; **Tycko, D. H., Thomas, L. H.**, and **King, K. M.**, *Phys. Rev.*, 109, 369, 1958.
64. For Be refer to **Watson, R. E.**, *Phys. Rev.*, 119, 170, 1960; **Bunge, C. F.**, *Phys. Rev.*, 168, 92, 1968; For B refer to **Schaefer, H. F.** and **Harris, F. E.**, *Phys. Rev.*, 167, 67, 1968.
65. **Green, L. C., Kolchin, E. K.**, and **Johnson, N. C.**, *Phys. Rev.*, 139, A373, 1965.
66. **Wilson, W. S.** and **Lindsay, R. B.**, *Phys. Rev.*, 74, 681, 1935.

67. Hylleraas, E. A., *Z. Phys.*, 54, 347, 1929.
68. Kinoshita, T., *Phys. Rev.*, 105, 1490, 1957.
69. Scherr, C. W. and Knight, R. E., *Phys. Rev.*, 35, 436, 1963.
70. Pekeris, C. L., *Phys. Rev.*, 112, 1649, 1959. In this ref. a series expansion in Laguerre Polynomials has been used instead of Hyllerass-type functions.
71. *Handbook of Chemistry and Physics*, 51st ed., Weast, R. C., Ed., CRC Press, Boca Raton, Fla., 1970.
72. *American Institute of Physics Handbook*, 2nd ed., McGraw-Hill, New York, 1963, Table 7C-1, 7.
73. Sternheimer, R. M., *Phys. Rev.*, A20, 18, 1979.
74. Sternheimer, R. M., *Phys. Rev.*, A19, 474, 1979.
75. Sternheimer, R. M., *Phys. Rev.*, A16, 1752, 1977.
76. Sternheimer, R. M., *Phys. Rev.*, A16, 459, 1977.
77. Sternheimer, R. M., *Phys. Rev.*, A15, 1918, 1977.
78. Hartree, D. R., *Proc. Cambridge Philos. Soc.*, 24, 89, 1928; 24, 111, 1928.
79. Hartree, D. R., *The Calculations of Atomic Structure*, John Wiley & Sons, New York, 1957.
80. For the suggestion of the variational method see also Slater, J. C., *Phys. Rev.*, 35, 210, 1930.
81. Fock, V., *Z. Phys.*, 61, 126, 1930; a general scheme was also suggested by Slater, J. C., *Phys. Rev.*, 35, 210, 1930. For the first calculation using antisymmetrized wavefunction see Fock, V., *Z. Phys.*, 62, 795, 1930; for further continuation also see Fock, V. and Petrashen, M., *Phys. Z. Sowjetunion*, 8, 359, 547, 1935.
82. Koopman, T. A., *Physica*, 1, 104, 1933.
83. For the compilation of the table of the Hartree-Fock radial functions refer to Landolt-Boernstein, *Zahlenwertz and Funktionen*, Vol. 1 (Part 1), Springer-Verlag, Berlin, 1950, 276.
84. Roothaan, C. C. J., *Rev. Modern Phys.*, 23, 69, 1951.
85. Synek, M., Raines, A. E., and Roothaan, C. C. J., *Phys. Rev.*, 141, 174, 1966.
86. Watson, R. E., *Phys. Rev.*, 118, 1036, 1960.
87. Watson, R. E. and Freeman, A. J., *Phys. Rev.*, 123, 521, 1961; 124, 1117, 1961.
88. Freeman, A. J. and Watson, R. E., *Phys. Rev.*, 127, 2058, 1962.
89. Piper, W. W., *Phys. Rev.*, 123, 1281, 1961.
90. Synek, M., *Phys. Rev.*, 139, A1049, 1965; 133, A961, 1964.
91. Roothaan, C. C. J. and Synek, M., *Phys. Rev.*, 133, A1263, 1964.
92. Lenander, C. J., *Phys. Rev.*, 130, 1033, 1963.
93. Cohen, S., *Phys. Rev.*, 118, 489, 1960.
94. Synek, M., Phys. Rev., 136, A112, 1964; Synek, M. and Stungis, G. E., *Phys. Rev.*, 136, 112, 1964.
95. Hartmann, H. and Clementi, E., *Phys. Rev.*, 133, A1295, 1964.
96. Waber, J. T. and Crommer, D. T., *J. Chem. Phys.*, 42, 4116, 1965.
97. Liberman, D., Waber, J. T., and Cromer, D. T., *Phys. Rev.*, 137, A27, 1965.
98. Kim, V., *Phys. Rev.*, 154, 17, 1967.
99. Clementi, E. and Raimondi, D. L., *J. Chem. Phys.*, 38, 2686, 1963.
100. Clementi, E., *J. Chem. Phys.*, 40, 1964, 1964; Clementi, E., *IBM J. Res. Develop.*, 9, 2, 1965.
101. Sharma, R. R., *Phys. Rev.*, A18, 726, 1978; Sharma, R. R. and Kolman, R., *J. Chem. Phys.*, 68, 2516, 1978; Sharma, R. R. and Moutsos, P., *Phys. Rev.*, B11, 1840, 1975; Moutsos, P., Adams, J., and Sharma, R. R., *J. Chem. Phys.*, 60, 1447, 1974.
102. Sharma, R. R., *Phys. Rev. Lett.*, 25, 1622, 1970; 26, 563, 1971; Sharma, R. R. and Teng, B. N., *Phys. Rev. Lett.*, 27, 679, 1971; Sharma, R. R. and Sharma, A. K., *Phys. Rev. Lett.*, 29, 122, 1972; Sharma, R. R., *Phys. Rev.*, B6, 4310, 1972.
103. Bagus, P. S., Gilbert, T. L., and Roothaan, C. C. J., Argonne National Laboratory Report, February, 1972.
104. Thomas, L. H., *Proc. Cambridge Philos. Soc.*, 23, 542, 1927.
105. Fermi, E., *Mem. Accad. Lincei*, 6, 602, 1927; *Collected Papers*, University of Chicago Press, Chicago, 1962; Gombas, P., *Rev. Mod. Phys.*, 35, 512, 1963; *Handbook of Physics*, Vol. 36, (Part 2), Flugge, S., Ed., Springer, Berlin, 1956, 108; March, N. H., *Adv. Phys.*, 6, 1, 1957.
106. Bates, D. A. and Damgaard, A., *Philos. Trans. R. Soc. London*, 242, 101, 1949.
107. Sigiura, Y., *Philos. Mag.*, 4, 498, 1927.
108. Armstrong, B. H. and Purdum, K. L., *Phys. Rev.*, 150, 51, 1966.
109. Slater, J. C., *Phys. Rev.*, 81, 385, 1951.
110. Maslen, V. W., *Proc. Phys. Soc. (London)*, A69, 734, 1956.
111. Slater, J. C., *Quantum Theory of Atomic Structure*, Vol. 1 and 2, McGraw-Hill, New York, 1960.
112. Herman, F., Callaway, J., and Acton, F. S., *Phys. Rev.*, 95, 371, 1954.
113. Hartree, D. R., *Phys. Rev.*, 109, 840, 1958.
114. Raimes, S. *The Wave Mechanics of Electrons in Metals*, Interscience, New York, 1961.
115. Robinson, J. E., Bassani, F., Knox, R. S., and Schrieffer, J. R., *Phys. Rev. Lett.*, 9, 215, 1962.

116. Prett, G. W., Jr., *Phys. Rev.*, 102, 1303, 1956.
117. Watson, R. E., *Phys. Rev.*, 118, 1036, 1960.
118. Watson, R. E., *Phys. Rev.*, 119, 1934, 1960.
119. Watson, R. E. and Freeman, A. J., *Phys. Rev.*, 123, 521, 2027, 1961.
120. Watson, R. E. and Freeman, A. J., *Phys. Rev.*, 120, 1125, 1134, 1960.
121. Watson, R. E. and Freeman, A. J., *Phys. Rev.* 124, 1117, 1961.
122. Goodings, D. A., *Phys. Rev.*, 123, 1706, 1961.
123. Latter, R., *Phys. Rev.*, 99, 510, 1955.
124. Herman, F. and Skillman, S., *Atomic Structure Calculations*, Prentice-Hall, Englewood Cliffs, N. J., 1963.
125. Hartree, D. R., *Proc. R. Soc. (London)*, A141, 282, 1933.
126. Hartree, D. R. and Hartree, W., *Proc. R. Soc. (London)*, A157, 490, 1936.
127. Slater, J. C., *The Self-Consistent Field for Molecules and Solids*, Vol. 4, McGraw-Hill, New York, 1974.
128. Slater, J. C. and Johnson, K. H., *Phys. Rev.*, B5, 844, 1972.
129. Kohn, W. and Sham, L. J., *Phys. Rev.*, 140, A1133, 1965.
130. Hohenberg, P. and Kohn, W., *Phys. Rev.*, 136, B864, 1964.
131. Stoddart, J. C. and March, N. H., *Ann. Phys. (N.Y.)*, 64, 174, 1971.
132. van Barth, U. and Hedin, L., *J. Phys.*, C5, 1629, 1972.
133. Rajagopal, A. K. and Callaway, J., *Phys. Rev.*, B7, 1912, 1973.
134. Gunnarsson, O. and Lundquist, B. I., *Phys. Rev.*, B13, 4274, 1976. An earlier account related to this work is given by Gunnarsson, O., Lundquist, B. I., and Lundquist, S., *Solid State Commun.*, 11, 149, 1972.
135. Gunnarsson, O. and Johansson, P., *Int. J. Quan. Chem.*, 10, 307, 1976.
136. For calculations on V. Fe, Co, Ni, Pd and Pt, see Gunnarsson, O., *J. Phys. F*, to be published.
137. Gunnarsson, O., Lundquist, B. I., and Wilkins, J. W., *Phys. Rev.*, B10, 1319, 1974.
138. Tong, B. Y., *Phys. Rev.*, B6, 1189, 1972.
139. Janak, J. F., Moruzzi, V. L., and Williams, A. R., *Phys. Rev.*, to be published.
140. Ekin, J. W. and Maxfield, B. W., *Phys. Rev.*, B4, 4215, 1971.
141. Popovic, Z. D. and Stott, M. J., *Phys. Rev. Lett.*, 33, 1164, 1974.
142. Lang, N. D., *Solid State Phys.*, 28, 225, 1973.
143. Borisov, Yu. A., *Chem. Phys. Lett.*, 44, 17, 1976.
144. For local density theory of multiplet structure, consult Van Barth, U., *Phys. Rev.*, A20, 1693, 1979.
145. For density functionals in unrestricted Hartree-Fock theory, see Payne, P. W., *J. Chem. Phys.*, 71, 490, 1979.
146. Lawes, G. P. and March, N. H., *J. Chem. Phys.*, 71, 1007, 1979. This deals with exact local density method for linear harmonic oscillator.
147. For a correlation between the trends in measured energies for 3d atoms and the calculated values using spin-density functional approach one may consult Harris, J. and Jones, R. O., *J. Chem. Phys.*, 68, 3316, 1978.
148. MacDonald, A. H. and Vosko, S. H., *J. Phys.*, C12, 2977, 1979. This deals with a relativistic generalization of the Hohenberg-Kohn-Sham density functional formalism.
149. Hellman, H. *Acta Fizichim. U.S.S.R.*, 1, 139, 1935.
150. Gombas, P., *Z. Phys.*, 118, 164, 1941.
151. Fock, V., Vesselow, M., and Petrashen, M., *Zh. Eksp. Theor. Fiz.*, 10, 723, 1940.
152. Abarenkov, I. V. and Heine, V., *Philos. Mag.*, 12, 529, 1965.
153. Schwarz, W. H., *Theor. Chim. Acta*, 11, 377, 1968.
154. Roach, A. C. and Child, M. S., *Mol. Phys.*, 14, 1, 1968.
155. Kutzelnigg, W., Koch, R. J., and Bingle, W. A., *Chem. Phys. Lett.*, 2, 197, 1968.
156. Simmons, G., *J. Chem. Phys.*, 55, 756, 1971.
157. Szasg, L. and McGinn, G., *J. Chem. Phys.*, 42, 2363, 1965.
158. Szasg, L. and McGinn, G., *J. Chem. Phys.*, 45, 2898, 1966.
159. Simons, G. and Mazziotti, A., *J. Chem. Phys.*, 52, 2449, 1970.
160. Goodfriend, P. L. and Hart, G. A., *J. Mol. Spectrosc.*, 42, 312, 1972.
161. Hart, G. A. and Goodfriend, P. L., *Mol. Phys.*, 29, 1109, 1975.
162. Schwartz, M. E. and Switalski, J. D., *J. Chem. Phys.*, 57, 4125, 4132, 1972.
163. Chang, T. C., Habitz, P., Pittel, B., and Schwartz, W. H. E., *Theor. Chim. Acta*, 34, 263, 1974.
164. Phillips, J. C. and Kleinmann, L., *Phys. Rev.*, 116, 287, 1959.
165. Weeks, J. D., Hazi, A., and Rice, S. A., *Adv. Chem. Phys.*, 16, 283, 1969.
166. Harrison, W. A., *Pseudopotentials in the Theory and Metals*, W. A. Benjamin, New York, 1966; Harrison, W. A., *Solid State Theory*, McGraw-Hill, New York, 1970.
167. Kahn, L. R., Baybutt, P., and Truhlar, D. G., *J. Chem. Phys.*, 65, 3826, 1976.

168. Coffey, P., Ewig, C. S., and Van Wazer, J. R., *J. Am. Chem. Soc.*, 97, 1656, 1975.
169. Bonifacic, V. and Huzinaga, S., *J. Chem. Phys.*, 60, 2779, 1974; *J. Chem. Phys.*, 64, 956, 1976.
170. Kahn, L. R. and Goddard, W. A., III, *J. Chem. Phys.*, 56, 2685, 1972.
171. Melius, C. F. and Goddard, W. A., III, *Phys. Rev.*, A10, 1528, 1974.
172. Melius, C. F., Olafson, B. D., and Goddard, W. A., III, *Chem. Phys. Lett.*, 28, 457, 1974.
173. Durand, P. and Barthelat, J. C., *Theor. Chim. Acta*, 38, 283, 1975.
174. Tiopol, S., Moskowitz, J. W., and Melius, C. F., *J. Chem. Phys.*, 68, 2364, 1975.
175. Hay, P. J., Wadt, W. R., and Kahn, L. R., *J. Chem. Phys.*, 68, 3059, 1978.
176. Harris, J. and Jones, R. O., *Phys. Rev. Lett.*, 41, 191, 1978.
177. Miller, D. J., Haneman, D., Baerends, E. J., and Ros, P., *Phys. Rev. Lett.*, 41, 197, 1978.
178. Wendel, H. and Martin, R. M., *Phys. Rev. Lett.*, 40, 950, 1978.
179. Stoll, H., Wagenblast, G., and Preuss, H., *Theor. Chim. Acta (Berl.)*, 49, 67, 1978; Stoll, H. and Preuss, H., *Theor. Chim. Acta (Berl.)*, 46, 11, 1977.
180. Hamann, D. R., Schlüter, M., and Chiang, C., *Phys. Rev. Lett.*, 43, 1494, 1979.
181. Appelbaum, J. A. and Hamann, D. R., *Phys. Rev.*, B8, 1777, 1973.
182. Kimura, K., *Phys. Lett.*, 72A, 456, 1979.
183. Gasper, R., Jr. and Gasper, R., *Int. J. Quant. Chem.*, 16, 57, 1979.
184. Chelikowsky, J. R. and Cohen, M. L., *Phys. Rev.*, B14, 556, 1976.
185. Goodfriend, P. L., *Am. J. Phys.*, 47, 630, 1979.
186. Moseley, H. G. J., *Philos. Mag.*, 26, 1024, 1913; *Philos. Mag.*, 27, 703, 1914.
187. Allen, J. M., *Phys. Rev.*, 27, 266, 1926; *Phys. Rev.*, 28, 907, 1926.
188. Leighton, R. B., *Principles of Modern Physics*, Figure 12-12, McGraw-Hill, New York, 1959.
189. de Broglie, L., *C. R.*, 158, 1493, 1914; *C. R.*, 163, 87, 1916; *C. R.*, 163, 352, 1916.
190. Wapstra, A. H., Nijgh, G. J., and Van Lieshout, R., *Nuclear Spectroscopy Tables*, North-Holland, New York, 1959.
191. Williams, J. H., *Phys. Rev.*, 44, 146, 1933.
192. Meyers, H. T., *Wiss. Verh. Siemens*, 7, 108, 1929.
193. Beckman, O., *Ark. Fys.*, 9, 495, 1955.
194. Beckman, O., *Phys. Rev.*, 109, 1590, 1958.
195. Gibbs, R. C., Wilber, D. T., and White, H. E., *Phys. Rev.*, 29, 790, 1927.
196. Moore, C. E., Atomic Energy Levels, National Bureau of Standards, Washington, D.C., 1952.
197. Weiss, W. L., Smith, M. W., and Glennon, B. M., Vol. 1 and 2, National Bureau of Standards, Washington, D. C., 1966.
198. Auger, P., *C. R.*, 180, 65, 1925; *C. R.*, 182, 773, 1215, 1926; *J. Phys. Radium*, 6, 205, 1925.
199. Bergstrom, I. and Nordling, C., in *Alpha-, Beta- and Gamma-Spectroscopy*, Siegbahn, K., Ed., North-Holland, Amsterdam, 1964.
200. Asaad, W. N., *Proc. R. Soc. (London)*, A249, 555, 1959.
201. Asaad, W. N., *Nucl. Phys.*, 66, 494, 1965; *Nucl. Phys.*, 63, 337, 1965.
202. Listengarten, M. A., *Izv. Akad. Nauk S.S.S.R. Ser. Fiz.*, 24, 1041, 1960; *Bull. Acad. Sci. U.S.S.R. Phys.*, 24, 1050, 1960.
203. Listengarten, M. A., *Izv. Akad. Nauk S.S.S.R. Ser. Fiz.*, 26, 182, 1962; *Bull. Akad. Sci. U.S.S.R. Phys.*, (Columbia Tech. Transl.), 26, 182, 1962; *Izv. Akad. Nauk Sci. U.S.S.R. Phys.*, (Columbia Tech. Transl.), 25, 803, 1961.
204. Dionisio, J. S, *Ann. Phys. (Paris)*, 8, 747, 1963.
205. Callan, E. J., *Rev. Mod. Phys.*, 35, 524, 1963; *Bull. Am. Phys. Soc.*, 7, N416, 1962.
206. Coster, D. and Kronig, R., *Physica*, 2, 13, 1935.
207. Burhop, E. H. S., *The Auger Effect and other Radiationless Transitions*, Cambridge University Press, New York, 1952.
208. Parilis, E. C., The Auger Effect, (in Russ.), *Acad. Sci. Uzbek S.S.R.*, Tashkent, 1969.
209. Fink, R. W., Jopson, R. C., Mark, H., and Swift, C. D., *Rev. Mod. Phys.*, 38, 513, 1966.
210. Burhop, E. H. S. and Asaad, W. N., The Auger effect, in *Advances in Atomic and Molecular Physics*, Vol. 8, Academic Press, New York, 1972, 164.
211. McGuire, E. J., in *Atomic Inner Shell Processes*, Vol. 1, Crasemann, B., Ed., Academic Press, New York, 1975, 293.
212. Bambynek, W., Crasemann, B., Fink, R. W., Freund, H. U., Mark, H., Swift, C. D., Price, R. E., and Rao, P. V., *Rev. Mod. Phys.*, 44, 716, 1972.
213. Slatis, H., *Ark. Fys.*, 37, 25, 1968.
214. Krisciokaitis, R. J. and Haynes, S. K., *Nucl. Instrum. Methods*, 58, 309, 1968.
215. Korber, H. and Mehlorn, W., *Z. Phys.*, 191, 217, 1966.
216. Krause, M. O., *Phys. Lett.*, 19, 14, 1965.
217. Grahm, R. L., Ewan, G. T., and Geiger, J. S., *Nucl. Instrum. Methods*, 9, 245, 1960.
218. Hedgran, A., Siegbahn, K., and Svartholm, N., *Proc. Phys. Soc. London, Sect. A*, 63, 960, 1950.

219. Baird, Q. L., Nall, J. C., Haynes, S. K., and Hamilton, J. H., *Nucl. Instrum. Methods*, 16, 275, 1962.
220. Asaad, W. N., *Nucl. Phys.*, 63, 337, 1965.
221. Listengarten, M. A., *Bull. Acad. Sci. U.S.S.R. Phys. Ser.*, 26, 182, 1962.
222. Asaad, W. N. and Burhop, E. H. S., *Proc. Phys. Soc., London*, 71, 369, 1958.
223. Bloch, F., *Phys. Rev.*, 48, 187, 1935.
224. Aberg, T. and Utriainen, J., *Phys. Rev. Lett.*, 1346, 1969.
225. Wentzel, G., *Z. Phys.*, 43, 524, 1927.
226. Mott, N. F. and Massey, H. S. W., *The Theory of Atomic Collisions*, 3rd ed., Oxford University Press, New York, 1965.
227. Bathe, H. A., *Intermediate Quantum Mechanics*, W. A. Benjamin, New York, 1964.
228. Scofield, J. H., *Phys. Rev.*, 179, 9, 1969; see also the references cited therein for the calculations of the relativistic radiative transition rates.
229. Raskar, W. and Raffray, Y., *C. R. Acad. Sci. Ser. B.*, 265, 23, 1307, 1967.
230. Massey, H. S. W. and Burhop, E. H. S., *Proc. Cambridge Philos. Soc.*, 32, 461, 1936.
231. Massey, H. S. W. and Burhop, E. H. S., *Proc. R. Soc. Ser. A*, 153, 661, 1936.
232. Wapstra, A., Nijgh, G., and Van Lieushout, R., *Nuclear Spectroscopy Tables*, North-Holland, Amsterdam, 1959.
233. Listergarten, M. A., *Izv. Akad. Nauk S.S.S.R. Ser. Fiz.*, 24, 1041 1960; *Bull. Acad. Sci. U.S.S.R. Phys.*, 24, 1040, 1960.
234. Chen, M. H., Crasemann, B., and Kostroun, V. O., *Phys. Rev.*, A4, 1, 1971.
235. Walters, D. L. and Bhalla, C. P., *Phys. Rev.*, A3, 519, 1971.
236. Walters, D. L. and Bhalla, C. P., *Phys. Rev.*, A3, 1919, 1971.
237. McGuire, E. J., *Phys. Rev.*, 185, 1, 1969.
238. McGuire, E. J., *Phys. Rev.*, A2, 273, 1970.
239. Rubenstein, R. A., Ph.D. thesis, University of Illinois, Urbana, 1955.
240. Mehlhorn, W., *Habilitationsschrift*, University of Münster, Federal Republic of Germany, 1967.
241. Kostroun, V. O., Chen, M. H. and Crasemann, B., *Phys. Rev.*, A3, 533, 1971.
242. Callen, E. J., Krueger, T. K., and McDavid, W. L., *Bull. Am. Phys. Soc.*, 14, 830, 1969.
243. Callen, E. J., *Phys. Rev.*, 124, 793, 1961.
244. Ramberg, E. and Richtmyer, F., *Phys. Rev.*, 51, 913, 1937.
245. Gaffrion, C. and Nadeu, G., Rep. TR 59—145, U.S. Air Force Off. Sci. Res., U.S. Department of Defense, Washington, D.C., 1959.
246. Pincherle, L., *Nuovo Cimento*, 12, 8, 1935.
247. Burhop, E. H. S., *Proc. R. Soc. Ser. A*, 148, 272, 1935.
248. McGuire, E. J., *Phys. Rev.*, A3, 587, 1971.
249. Chen, M. H., Laiman, E., and Crasemann, B., *Phys. Rev.*, A19, 2253, 1979.
250. Bhalla, C. P., *J. Phys.*, B3, L9, 1970; *Phys.*, A2, 722, 1970; *J. Phys.*, B3, 916, 1970; Bhalla, C. P. and Ramsdale, D. J., *J. Phys.* B, 3, L14, 1970.
251. Chatterji, D. and Talukdar, B., *Phys. Rev.*, 174, 44, 1968.

INDEX

A

B

C

D

H

I

J

K

L

M

N

O

P

Q

R

S

T

U

V

W

X

Y

Z

Printed and bound by CPI Group (UK) Ltd, Croydon, CR0 4YY
22/10/2024
01777637-0010